Introduction

Mathematical Thinking at Grade 3

Grade 3

Also appropriate for Grade 4

Susan Jo Russell
Karen Economopoulos

Developed at TERC, Cambridge, Massachusetts

Dale Seymour Publications®

The *Investigations* curriculum was developed at TERC (formerly Technical Education Research Centers) in collaboration with Kent State University and the State University of New York at Buffalo. The work was supported in part by National Science Foundation Grant No. MDR-9050210. TERC is a nonprofit company working to improve mathematics and science education. TERC is located at 2067 Massachusetts Avenue, Cambridge, MA 02140.

This project was supported, in part, by the

National Science Foundation

Opinions expressed are those of the authors and not necessarily those of the Foundation

This book is published by Dale Seymour Publications®, an imprint of Addison Wesley Longman, Inc.

Managing Editor: Catherine Anderson
Series Editor: Beverly Cory
Revision Team: Laura Marshall Alavosus, Ellen Harding, Patty Green Holubar, Suzanne Knott, Beverly Hersh Lozoff
ESL Consultant: Nancy Sokol Green
Production/Manufacturing Director: Janet Yearian
Production/Manufacturing Coordinator: Barbara Atmore
Design Manager: Jeff Kelly
Design: Don Taka
Illustrations: Susan Jaekel, Carl Yoshihara
Cover: Bay Graphics
Composition: Archetype Book Composition

 Printed on Recycled Paper

Order number DS43841
ISBN 1-57232-694-8
1 2 3 4 5 6 7 8 9 10-ML-01 00 99 98 97

T E R C

Principal Investigator   Susan Jo Russell

Co-Principal Investigator   Cornelia C. Tierney

Director of Research and Evaluation   Jan Mokros

Curriculum Development
Joan Akers
Michael T. Battista
Mary Berle-Carman
Douglas H. Clements
Karen Economopoulos
Ricardo Nemirovsky
Andee Rubin
Susan Jo Russell
Cornelia C. Tierney
Amy Shulman Weinberg

Evaluation and Assessment
Mary Berle-Carman
Abouali Farmanfarmaian
Jan Mokros
Mark Ogonowski
Amy Shulman Weinberg
Tracey Wright
Lisa Yaffee

Teacher Support
Rebecca B. Corwin
Karen Economopoulos
Tracey Wright
Lisa Yaffee

Technology Development
Michael T. Battista
Douglas H. Clements
Julie Sarama Meredith
Andee Rubin

Video Production
David A. Smith

Administration and Production
Amy Catlin
Amy Taber

*Cooperating Classrooms
for This Unit*
Stephanie Burgess
Corrine Varon
Virginia M. Micciche
Cambridge Public Schools
Cambridge, MA

Jeanne Wall
Caroline Thom
Arlington Public Schools
Arlington, MA

Katie Bloomfield
Robert A. Dihlmann
Shutesbury Elementary
Shutesbury, MA

Consultants and Advisors
Elizabeth Badger
Deborah Lowenberg Ball
Marilyn Burns
Ann Grady
Joanne M. Gurry
James J. Kaput
Steven Leinwand
Mary M. Lindquist
David S. Moore
John Olive
Leslie P. Steffe
Peter Sullivan
Grayson Wheatley
Virginia Woolley
Anne Zarinnia

Graduate Assistants
Kent State University
Joanne Caniglia
Pam DeLong
Carol King

State University of New York at Buffalo
Rosa Gonzalez
Sue McMillen
Julie Sarama Meredith
Sudha Swaminathan

Revisions and Home Materials
Cathy Miles Grant
Marlene Kliman
Margaret McGaffigan
Megan Murray
Kim O'Neil
Andee Rubin
Susan Jo Russell
Lisa Seyferth
Myriam Steinback
Judy Storeyguard
Anna Suarez
Cornelia Tierney
Carol Walker
Tracey Wright

CONTENTS

TEACHER NOTES

WHERE TO START

The first-time user of *Mathematical Thinking at Grade 3* should read the following:

When you next teach this same unit, you can begin to read more of the background. Each time you present the unit, you will learn more about how your students understand the mathematical ideas.

Investigations in Number, Data, and Space® is a K–5 mathematics curriculum with four major goals:

- to offer students meaningful mathematical problems
- to emphasize depth in mathematical thinking rather than superficial exposure to a series of fragmented topics
- to communicate mathematics content and pedagogy to teachers
- to substantially expand the pool of mathematically literate students

The *Investigations* curriculum embodies a new approach based on years of research about how children learn mathematics. Each grade level consists of a set of separate units, each offering 2–8 weeks of work. These units of study are presented through investigations that involve students in the exploration of major mathematical ideas.

Approaching the mathematics content through investigations helps students develop flexibility and confidence in approaching problems, fluency in using mathematical skills and tools to solve problems, and proficiency in evaluating their solutions. Students also build a repertoire of ways to communicate about their mathematical thinking, while their enjoyment and appreciation of mathematics grows.

The investigations are carefully designed to invite all students into mathematics—girls and boys, members of diverse cultural, ethnic, and language groups, and students with different strengths and interests. Problem contexts often call on students to share experiences from their family, culture, or community. The curriculum eliminates barriers—such as work in isolation from peers, or emphasis on speed and memorization—that exclude some students from participating successfully in mathematics. The following aspects of the curriculum ensure that all students are included in significant mathematics learning:

- Students spend time exploring problems in depth.
- They find more than one solution to many of the problems they work on.

- They invent their own strategies and approaches, rather than relying on memorized procedures.
- They choose from a variety of concrete materials and appropriate technology, including calculators, as a natural part of their everyday mathematical work.
- They express their mathematical thinking through drawing, writing, and talking.
- They work in a variety of groupings—as a whole class, individually, in pairs, and in small groups.
- They move around the classroom as they explore the mathematics in their environment and talk with their peers.

While reading and other language activities are typically given a great deal of time and emphasis in elementary classrooms, mathematics often does not get the time it needs. If students are to experience mathematics in depth, they must have enough time to become engaged in real mathematical problems. We believe that a minimum of five hours of mathematics classroom time a week—about an hour a day—is critical at the elementary level. The plan and pacing of the *Investigations* curriculum is based on that belief.

We explain more about the pedagogy and principles that underlie these investigations in Teacher Notes throughout the units. For correlations of the curriculum to the NCTM Standards and further help in using this research-based program for teaching mathematics, see the following books:

- *Implementing the* Investigations in Number, Data, and Space® *Curriculum*
- *Beyond Arithmetic: Changing Mathematics in the Elementary Classroom* by Jan Mokros, Susan Jo Russell, and Karen Economopoulos

This book is one of the curriculum units for *Investigations in Number, Data, and Space.* In addition to providing part of a complete mathematics curriculum for your students, this unit offers information to support your own professional development. You, the teacher, are the person who will make this curriculum come alive in the classroom; the book for each unit is your main support system.

Although the curriculum does not include student textbooks, reproducible sheets for student work are provided in the unit and are also available as Student Activity Booklets. Students work actively with objects and experiences in their own environment and with a variety of manipulative materials and technology, rather than with a book of instruction and problems. We strongly recommend use of the overhead projector as a way to present problems, to focus group discussion, and to help students share ideas and strategies.

Ultimately, every teacher will use these investigations in ways that make sense for his or her particular style, the particular group of students, and the constraints and supports of a particular school environment. Each unit offers information and guidance for a wide variety of situations, drawn from our collaborations with many teachers and students over many years. Our goal in this book is to help you, a professional educator, implement this curriculum in a way that will give all your students access to mathematical power.

Investigation Format

The opening two pages of each investigation help you get ready for the work that follows.

What Happens This gives a synopsis of each session or block of sessions.

Mathematical Emphasis This lists the most important ideas and processes students will encounter in this investigation.

What to Plan Ahead of Time These lists alert you to materials to gather, sheets to duplicate, transparencies to make, and anything else you need to do before starting.

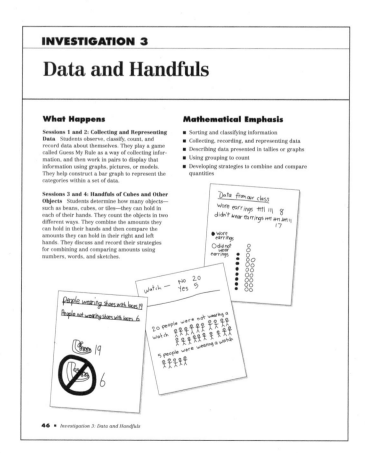

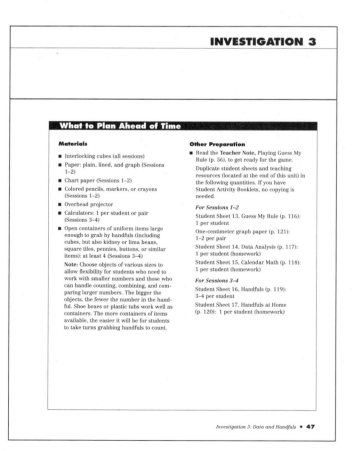

Sessions Within an investigation, the activities are organized by class session, a session being at least a one-hour math class. Sessions are numbered consecutively through an investigation. Often several sessions are grouped together, presenting a block of activities with a single major focus.

When you find a block of sessions presented together—for example, Sessions 1, 2, and 3—read through the entire block first to understand the overall flow and sequence of the activities. Make some preliminary decisions about how you will divide the activities into three sessions for your class, based on what you know about your students. You may need to modify your initial plans as you progress through the activities, and you may want to make notes in the margins of the pages as reminders for the next time you use the unit.

Be sure to read the Session Follow-Up section at the end of the session block to see what homework assignments and extensions are suggested as you make your initial plans.

While you may be used to a curriculum that tells you exactly what each class session should cover, we have found that the teacher is in a better position to make these decisions. Each unit is flexible and may be handled somewhat differently by every teacher. While we provide guidance for how many sessions a particular group of activities is likely to need, we want you to be active in determining an appropriate pace and the best transition points for your class. It is not unusual for a teacher to spend more or less time than is proposed for the activities.

Ten-Minute Math At the beginning of some sessions, you will find Ten-Minute Math activities. These are designed to be used in tandem with the investigations, but not during the math hour. Rather, we hope you will do them whenever you have a spare 10 minutes—maybe before lunch or recess, or at the end of the day.

Ten-Minute Math offers practice in key concepts, but not always those being covered in the unit. For example, in a unit on using data, Ten-Minute Math might revisit geometric activities done earlier in the year. Complete directions for the suggested activities are included at the end of each unit.

Sessions 1 and 2

Collecting and Representing Data

Materials

- Interlocking cubes
- Paper: plain, lined, and graph
- Chart paper
- Colored pencils, markers, or crayons
- Student Sheet 13 (1 per pair)
- Student Sheet 14 (1 per student, homework)
- Student Sheet 15 (1 per student, homework)
- Overhead projector

What Happens

Students observe, classify, count, and record data about themselves. They play a game called Guess My Rule as a way of collecting information and then work in pairs to display that information using graphs, pictures, or models. They help construct a bar graph to represent the categories within a set of data. Their work focuses on:

- collecting information about a group of people
- sorting and classifying information
- counting and comparing sets of data
- using pictures, tallies, and graphs to organize and display data

Activity

Playing Guess My Rule

Introduce this investigation by telling students that, as part of their work in mathematics this year, they will sometimes collect information about themselves and their families or about groups of people in the school. They will figure out ways to organize and describe the information they collect. One of the ways they will organize information is by thinking about the way things can be grouped.

When scientists and mathematicians study the world, they often try to think about how things are the same and different. Sometimes things go together one way, but if you think about them differently, you'll see they can go together in a new way.

For example, some people think all third graders go together because they are in the third grade, and some people think certain third graders go with certain second graders or fourth graders because they all play baseball, all read the same kind of books, or all walk to school.

Focus on characteristics of students in your classroom to give other examples about how students might go together in different ways.

Activities The activities include pair and small-group work, individual tasks, and whole-class discussions. In any case, students are seated together, talking and sharing ideas during all work times. Students most often work cooperatively, although each student may record work individually.

Choice Time In some units, some sessions are structured with activity choices. In these cases, students may work simultaneously on different activities focused on the same mathematical ideas. Students choose which activities they want to do, and they cycle through them.

You will need to decide how to set up and introduce these activities and how to let students make their choices. Some teachers present them as station activities, in different parts of the room. Some list the choices on the board as reminders or have students keep their own lists.

Extensions Sometimes in Session Follow-Up, you will find suggested extension activities. These are opportunities for some or all students to explore a topic in greater depth or in a different context. They are not designed for "fast" students; mathematics is a multifaceted discipline, and different students will want to go further in different investigations. Look for and encourage the sparks of interest and enthusiasm you see in your students, and use the extensions to help them pursue these interests.

Excursions Some of the *Investigations* units include excursions—blocks of activities that could be omitted without harming the integrity of the unit. This is one way of dealing with the great depth and variety of elementary mathematics— much more than a class has time to explore in any one year. Excursions give you the flexibility to make different choices from year to year, doing the excursion in one unit this time, and next year trying another excursion.

Tips for the Linguistically Diverse Classroom At strategic points in each unit, you will find concrete suggestions for simple modifications of the teaching strategies to encourage the participation of all students. Many of these tips offer alternative ways to elicit critical thinking from students at varying levels of English proficiency, as well as from other students who find it difficult to verbalize their thinking.

The tips are supported by suggestions for specific vocabulary work to help ensure that all students can participate fully in the investigations. The Preview for the Linguistically Diverse Classroom (p. I-22) lists important words that are assumed as part of the working vocabulary of the unit. Second-language learners will need to become familiar with these words in order to understand the problems and activities they will be doing. These terms can be incorporated into students' second-language work before or during the unit. Activities that can be used to present the words are found in the appendix, Vocabulary Support for Second-Language Learners (p. 91). In addition, ideas for making connections to students' language and cultures, included on the Preview page, help the class explore the unit's concepts from a multicultural perspective.

Materials

A complete list of the materials needed for teaching this unit is found on p. I-17. Some of these materials are available in kits for the *Investigations* curriculum. Individual items can also be purchased from school supply dealers.

Classroom Materials In an active mathematics classroom, certain basic materials should be available at all times: interlocking cubes, pencils, unlined paper, graph paper, calculators, things to count with, and measuring tools. Some activities in this curriculum require scissors and glue sticks or tape. Stick-on notes and large paper are also useful materials throughout.

So that students can independently get what they need at any time, they should know where these materials are kept, how they are stored, and how they are to be returned to the storage area. For example, interlocking cubes are best stored in towers of ten; then, whatever the activity, they should be returned to storage in groups of ten at the end of the hour. You'll find that establishing such routines at the beginning of the year is well worth the time and effort.

Technology Calculators are used throughout *Investigations*. Many of the units recommend that you have at least one calculator for each pair. You will find calculator activities, plus Teacher Notes discussing this important mathematical tool, in an early unit at each grade level. It is assumed that calculators will be readily available for student use.

Computer activities at grade 3 use two software programs that were developed especially for the *Investigations* curriculum. *Tumbling Tetrominoes* is introduced in the 2-D Geometry unit, *Flips, Turns, and Area.* This game emphasizes ideas about area and about geometric motions (slides, flips, and turns). The program *Geo-Logo*™ is introduced in a second 2-D Geometry unit, *Turtle Paths,* where students use it to explore geometric shapes.

How you use the computer activities depends on the number of computers you have available. Suggestions are offered in the geometry units for how to organize different types of computer environments.

Children's Literature Each unit offers a list of suggested children's literature (p. I-17) that can be used to support the mathematical ideas in the unit. Sometimes an activity is based on a specific children's book, with suggestions for substitutions where practical. While such activities can be adapted and taught without the book, the literature offers a rich introduction and should be used whenever possible.

Student Sheets and Teaching Resources Student recording sheets and other teaching tools needed for both class and homework are provided as reproducible blackline masters at the end of each unit. They are also available as Student Activity Booklets. These booklets contain all the sheets each student will need for individual work, freeing you from extensive copying (although you may need or want to copy the occasional teaching resource on transparency film or card stock, or make extra copies of a student sheet).

We think it's important that students find their own ways of organizing and recording their work. They need to learn how to explain their thinking with both drawings and written words, and how to organize their results so someone else can under-

Name _____ Date _____
Student Sheet 3

Partially Filled 100 Chart

1	2			5		7			10
11					16				
			24					29	
	32						38		
41							48		
		53		55					60
		63				67			
	72		74		76			79	80
81							88		
				95		97			100

97   *Investigation 1 • Sessions 2–3*
Mathematical Thinking at Grade 3

stand them. For this reason, we deliberately do not provide student sheets for every activity. Regardless of the form in which students do their work, we recommend that they keep a mathematics notebook or folder so that their work is always available for reference.

Homework In *Investigations*, homework is an extension of classroom work. Sometimes it offers review and practice of work done in class, sometimes preparation for upcoming activities, and sometimes numerical practice that revisits work in earlier units. Homework plays a role both in supporting students' learning and in helping inform families about the ways in which students in this curriculum work with mathematical ideas.

Depending on your school's homework policies and your own judgment, you may want to assign more homework than is suggested in the units. For this purpose you might use the practice pages, included as blackline masters at the end of this unit, to give students additional work with numbers.

For some homework assignments, you will want to adapt the activity to meet the needs of a variety of students in your class: those with special needs, those ready for more challenge, and second-language learners. You might change the numbers in a problem, make the activity more or less complex, or go through a sample activity with those who need extra help. You can modify any student sheet for either homework or class use. In particular, making numbers in a problem smaller or larger can make the same basic activity appropriate for a wider range of students.

Another issue to consider is how to handle the homework that students bring back to class—how to recognize the work they have done at home without spending too much time on it. Some teachers hold a short group discussion of different approaches to the assignment; others ask students to share and discuss their work with a neighbor, or post the homework around the room and give students time to tour it briefly. If you want to keep track of homework students bring in, be sure it ends up in a designated place.

Investigations at **Home** It is a good idea to make your policy on homework explicit to both students and their families when you begin teaching with *Investigations*. How frequently will you be assigning homework? When do you expect homework to be completed and brought back to school? What are your goals in assigning homework? How independent should families expect their children to be? What should the parent or guardian's role be? The more explicit you can be about your expectations, the better the homework experience will be for everyone.

Investigations at Home (a booklet available separately for each unit, to send home with students) gives you a way to communicate with families about the work students are doing in class. This booklet includes a brief description of every session, a list of the mathematics content emphasized in each investigation, and a discussion of each homework assignment to help families more effectively support their children. Whether or not you are using the *Investigations* at Home booklets, we expect you to make your own choices about home-

work assignments. Feel free to omit any and to add extra ones you think are appropriate.

Family Letter A letter that you can send home to students' families is included with the blackline masters for each unit. Families need to be informed about the mathematics work in your classroom; they should be encouraged to participate in and support their children's work. A reminder to send home the letter for each unit appears in one of the early investigations. These letters are also available separately in Spanish, Vietnamese, Cantonese, Hmong, and Cambodian.

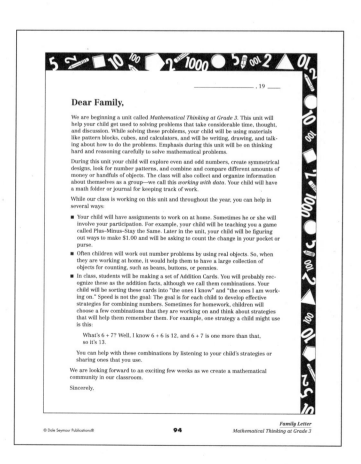

Help for You, the Teacher

Because we believe strongly that a new curriculum must help teachers think in new ways about mathematics and about their students' mathematical thinking processes, we have included a great deal of material to help you learn more about both.

About the Mathematics in This Unit This introductory section (p. I-18) summarizes the critical information about the mathematics you will be teaching. It describes the unit's central mathematical ideas and how students will encounter them through the unit's activities.

Teacher Notes These reference notes provide practical information about the mathematics you are teaching and about our experience with how students learn. Many of the notes were written in response to actual questions from teachers, or to discuss important things we saw happening in the field-test classrooms. Some teachers like to read them all before starting the unit, then review them as they come up in particular investigations.

Dialogue Boxes Sample dialogues demonstrate how students typically express their mathematical ideas, what issues and confusions arise in their thinking, and how some teachers have guided class discussions.

These dialogues are based on the extensive classroom testing of this curriculum; many are word-for-word transcriptions of recorded class discussions. They are not always easy reading; sometimes it may take some effort to unravel what the students are trying to say. But this is the value of these dialogues; they offer good clues to how your students may develop and express their approaches and strategies, helping you prepare for your own class discussions.

Where to Start You may not have time to read everything the first time you use this unit. As a first-time user, you will likely focus on understanding the activities and working them out with your students. Read completely through each investigation before starting to present it. Also read those sections listed in the Contents under the heading Where to Start (p. vi).

Teacher Note ▷   *Two Powerful Addition Strategies*

Most of us who are teaching today learned to add starting with the ones, then the tens, then the hundreds, and so on, moving from right to left and "carrying" from one column to another. This algorithm is certainly efficient once it is mastered. However, there are many other ways of adding that are just as efficient, closer to how we naturally think about quantities, connect better with good estimation strategies, and generally result in fewer errors.

When students rely on memorized rules and procedures that they do not understand, they usually do not estimate or double-check. They can easily make mistakes that make no sense. We want students to think about the quantities they are using and what results to expect. We want them to use their knowledge of the number system and important landmarks in that system. We want them to break apart and recombine numbers in ways that make computation more straightforward and, therefore, less prone to error. Writing problems horizontally rather than vertically is one way to help students focus on the whole quantities.

The two addition strategies discussed here are familiar to many competent users of mathematics. Your students may well invent others. Every student must be comfortable with more than one way of adding, so that an answer obtained using one method can be checked by using another. Anyone can make a mistake in routine computation, even with a calculator. What is critical, when accuracy matters, is that you have spent enough time estimating and double-checking to be able to rely on your result.

Left to Right Addition: Biggest Quantities First When students develop their own strategies for addition from an early age, they usually move from left to right, starting with the bigger parts of the quantities. For example, when adding 27 + 27, a student might say "20 and 20 is 40, then 7 and 7 is 14, so 40 plus 10 more makes 54." This strategy is both efficient and accurate. Some people who are

extremely good at computation use this strategy, even with large numbers.

One advantage of this approach is that when students work with the largest quantities first, it's easier to maintain a good sense of what the final sum should be. Another advantage is that students tend to continue seeing the two 27's as whole quantities, rather than breaking them up into their separate digits. When using the traditional algorithm (7 + 7 is 14, put down the 4, carry the 1), students too often see the two 7's, the 4, the 1, and the two 2's as individual digits. They lose their sense of the quantities involved, and if they end up with a nonsensical answer, they do not see it because they believe they "did it the right way."

Rounding to Nearby Landmarks Changing a number to a more familiar one that is easier to compute is another strategy that students should develop. Multiples of 10 and 100 are especially useful landmarks for students at this age. For example, to add 199 and 149, you might think of the problem as 200 plus 150, find the total of 350, then subtract 2 to compensate for the 2 you added on at the beginning.

To add 27 and 27, as in the previous example, some students might think of the problem as 30 + 30, then subtract 3 and 3 to give them the final result. Of course there are other useful landmarks, too. Another student might think of this problem as 25 + 25 + 2 + 2. There are no rules about which landmarks in the number system are best. It simply depends on whether using landmarks helps you solve the problem.

If your students have memorized the traditional right-to-left algorithm and believe that this is how they are "supposed" to do addition, you will have to work hard to instill some new values—that looking at the whole problem first and estimating the result is critical, that having more than one strategy is a necessary part of doing computation, and that using what you know about the numbers to simplify the problem leads to procedures that make sense.

D I A L O G U E   B O X

Playing Guess My Rule

This class is playing Guess My Rule (p. 48), and the teacher's secret rule is WEARING GREEN.

Su-Mei and Mark both fit the rule I'm thinking of. Let's have people who fit my rule stand here. [*Su-Mei and Mark stand up in front of the chalkboard.*] **Who thinks they know someone else who might fit this group? Don't guess the rule; just tell me another person you think might fit.**

Liliana: Do I fit?

Yes, you do fit my rule. [*Liliana joins Su-Mei and Mark. It happens that all three children have black hair. This characteristic is visually striking when they all stand together.*]

Maria [*who has black hair*]: I think I fit.

No, you don't fit the secret rule, but I bet I know what you were thinking about. Stand over by my desk to start the "people who don't fit the rule" group. Maria is an important clue.

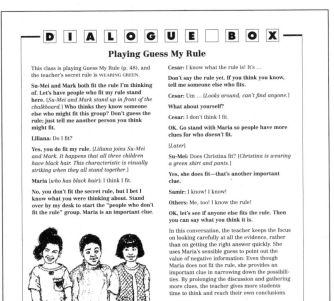

Cesar: I know what the rule is! It's ...

Don't say the rule yet. If you think you know, tell me someone else who fits.

Cesar: Um ... [*Looks around, can't find anyone.*]

What about yourself?

Cesar: I don't think I fit.

OK. Go stand with Maria so people have more clues for who *doesn't* fit.

[*Later*]

Su-Mei: Does Christina fit? [*Christina is wearing a green shirt and pants.*]

Yes, she does fit—that's another important clue.

Samir: I know! I know!

Others: Me, too! I know the rule!

OK, let's see if anyone else fits the rule. Then you can say what you think it is.

In this conversation, the teacher keeps the focus on looking carefully at all the evidence, rather than on getting the right answer quickly. She uses Maria's sensible guess to point out the value of negative information: Even though Maria does not fit the rule, she provides an important clue in narrowing down the possibilities. By prolonging the discussion and gathering more clues, the teacher gives more students time to think and reach their own conclusions.

The *Investigations* curriculum incorporates the use of two forms of technology in the classroom: calculators and computers. Calculators are assumed to be standard classroom materials, available for student use in any unit. Computers are explicitly linked to one or more units at each grade level; they are used with the unit on 2-D geometry unit at each grade, as well as with some of the units on measuring, data, and changes.

Using Calculators

In this curriculum, calculators are considered tools for doing mathematics, similar to pattern blocks or interlocking cubes. Just as with other tools, students must learn both *how* to use calculators correctly and *when* they are appropriate to use. This knowledge is crucial for daily life, as calculators are now a standard way of handling numerical operations, both at work and at home.

Using a calculator correctly is not a simple task; it depends on a good knowledge of the four operations and of the number system, so that students can select suitable calculations and also determine what a reasonable result would be. These skills are the basis of any work with numbers, whether or not a calculator is involved.

Unfortunately, calculators are often seen as tools to check computations with, as if other methods are somehow more fallible. Students need to understand that any computational method can be used to check any other; it's just as easy to make a mistake on the calculator as it is to make a mistake on paper or with mental arithmetic. Throughout this curriculum, we encourage students to solve computation problems in more than one way in order to double-check their accuracy. We present mental arithmetic, paper-and-pencil computation, and calculators as three possible approaches.

In this curriculum we also recognize that, despite their importance, calculators are not always appropriate in mathematics instruction. Like any tools, calculators are useful for some tasks, but not for others. You will need to make decisions about when to allow students access to calculators and when to ask that they solve problems without them, so that they can concentrate on other tools and skills. At times when calculators are or are not appropriate for a particular activity, we make specific recommendations. Help your students develop their own sense of which problems they can tackle with their own reasoning and which ones might be better solved with a combination of their own reasoning and the calculator.

Managing calculators in your classroom so that they are a tool, and not a distraction, requires some planning. When calculators are first introduced, students often want to use them for everything, even problems that can be solved quite simply by other methods. However, once the novelty wears off, students are just as interested in developing their own strategies, especially when these strategies are emphasized and valued in the classroom. Over time, students will come to recognize the ease and value of solving problems mentally, with paper and pencil, or with manipulatives, while also understanding the power of the calculator to facilitate work with larger numbers.

Experience shows that if calculators are available only occasionally, students become excited and distracted when they are permitted to use them. They focus on the tool rather than on the mathematics. In order to learn when calculators are appropriate and when they are not, students must have easy access to them and use them routinely in their work.

If you have a calculator for each student, and if you think your students can accept the responsibility, you might allow them to keep their calculators with the rest of their individual materials, at least for the first few weeks of school. Alternatively, you might store them in boxes on a shelf, number each calculator, and assign a corresponding number to each student. This system can give students a sense of ownership while also helping you keep track of the calculators.

Using Computers

Students can use computers to approach and visualize mathematical situations in new ways. The computer allows students to construct and manipulate geometric shapes, see objects move according to rules they specify, and turn, flip, and repeat a pattern.

This curriculum calls for computers in units where they are a particularly effective tool for learning mathematics content. One unit on 2-D geometry at each of the grades 3–5 includes a core of activities that rely on access to computers, either in the classroom or in a lab. Other units on geometry, measurement, data, and changes include computer activities, but can be taught without them. In these units, however, students' experience is greatly enhanced by computer use.

The following list outlines the recommended use of computers in this curriculum:

Grade 1
Unit: *Survey Questions and Secret Rules*
   (Collecting and Sorting Data)
Software: Tabletop, Jr.
Source: Broderbund

Unit: *Quilt Squares and Block Towns*
   (2-D and 3-D Geometry)
Software: *Shapes*
Source: provided with the unit

Grade 2
Unit: *Mathematical Thinking at Grade 2*
   (Introduction)
Software: *Shapes*
Source: provided with the unit

Unit: *Shapes, Halves, and Symmetry* (Geometry
   and Fractions)
Software: *Shapes*
Source: provided with the unit

Unit: *How Long? How Far?* (Measuring)
Software: *Geo-Logo*
Source: provided with the unit

Grade 3
Unit: *Flips, Turns, and Area* (2-D Geometry)
Software: *Tumbling Tetrominoes*
Source: provided with the unit

Unit: *Turtle Paths* (2-D Geometry)
Software: *Geo-Logo*
Source: provided with the unit

Grade 4
Unit: *Sunken Ships and Grid Patterns*
   (2-D Geometry)
Software: *Geo-Logo*
Source: provided with the unit

Grade 5
Unit: *Picturing Polygons* (2-D Geometry)
Software: *Geo-Logo*
Source: provided with the unit

Unit: *Patterns of Change* (Tables and Graphs)
Software: *Trips*
Source: provided with the unit

Unit: *Data: Kids, Cats, and Ads* (Statistics)
Software: Tabletop, Sr.
Source: Broderbund

The software provided with the *Investigations* units uses the power of the computer to help students explore mathematical ideas and relationships that cannot be explored in the same way with physical materials. With the *Shapes* (grades 1–2) and *Tumbling Tetrominoes* (grade 3) software, students explore symmetry, pattern, rotation and reflection, area, and characteristics of 2-D shapes. With the *Geo-Logo* software (grades 3–5), students investigate rotations and reflections, coordinate geometry, the properties of 2-D shapes, and angles. The *Trips* software (grade 5) is a mathematical exploration of motion in which students run experiments and interpret data presented in graphs and tables.

We suggest that students work in pairs on the computer; this not only maximizes computer resources but also encourages students to consult, monitor, and teach one another. Generally, more than two students at one computer find it difficult to share. Managing access to computers is an issue for every classroom. The curriculum gives you explicit support for setting up a system. The units are structured on the assumption that you have enough computers for half your students to work on the machines in pairs at one time. If you do not have access to that many computers, suggestions are made for structuring class time to use the unit with five to eight computers, or even with fewer than five.

Assessment plays a critical role in teaching and learning, and it is an integral part of the *Investigations* curriculum. For a teacher using these units, assessment is an ongoing process. You observe students' discussions and explanations of their strategies on a daily basis and examine their work as it evolves. While students are busy recording and representing their work, working on projects, sharing with partners, and playing mathematical games, you have many opportunities to observe their mathematical thinking. What you learn through observation guides your decisions about how to proceed. In any of the units, you will repeatedly consider questions like these:

- Do students come up with their own strategies for solving problems, or do they expect others to tell them what to do? What do their strategies reveal about their mathematical understanding?

- Do students understand that there are different strategies for solving problems? Do they articulate their strategies and try to understand other students' strategies?

- How effectively do students use materials as tools to help with their mathematical work?

- Do students have effective ideas for keeping track of and recording their work? Does keeping track of and recording their work seem difficult for them?

You will need to develop a comfortable and efficient system for recording and keeping track of your observations. Some teachers keep a clipboard handy and jot notes on a class list or on adhesive labels that are later transferred to student files. Others keep loose-leaf notebooks with a page for each student and make weekly notes about what they have observed in class.

Assessment Tools in the Unit

With the activities in each unit, you will find questions to guide your thinking while observing the students at work. You will also find two built-in assessment tools: Teacher Checkpoints and embedded Assessment activities.

Teacher Checkpoints The designated Teacher Checkpoints in each unit offer a time to "check in" with individual students, watch them at work, and ask questions that illuminate how they are thinking.

At first it may be hard to know what to look for, hard to know what kinds of questions to ask. Students may be reluctant to talk; they may not be accustomed to having the teacher ask them about their work, or they may not know how to explain their thinking. Two important ingredients of this process are asking students open-ended questions about their work and showing genuine interest in how they are approaching the task. When students see that you are interested in their thinking and are counting on them to come up with their own ways of solving problems, they may surprise you with the depth of their understanding.

Teacher Checkpoints also give you the chance to pause in the teaching sequence and reflect on how your class is doing overall. Think about whether you need to adjust your pacing: Are most students fluent with strategies for solving a particular kind of problem? Are they just starting to formulate good strategies? Or are they still struggling with how to start? Depending on what you see as the students work, you may want to spend more time on similar problems, change some of the problems to use smaller numbers, move quickly to more challenging material, modify subsequent activities for some students, work on particular ideas with a small group, or pair students who have good strategies with those who are having more difficulty.

Embedded Assessment Activities Assessment activities embedded in each unit will help you examine specific pieces of student work, figure out what it means, and provide feedback. From the students' point of view, these assessment activities are no different from any others. Each is a learning experience in and of itself, as well as an opportunity for you to gather evidence about students' mathematical understanding.

The embedded assessment activities sometimes involve writing and reflecting; at other times, a discussion or brief interaction between student and teacher; and in still other instances, the creation and explanation of a product. In most cases, the assessments require that students *show* what they did, *write* or *talk* about it, or do both. Having to explain how they worked through a problem helps students be more focused and clear in their mathematical thinking. It also helps them realize that doing mathematics is a process that may involve tentative starts, revising one's approach, taking different paths, and working through ideas.

Teachers often find the hardest part of assessment to be interpreting their students' work. We provide guidelines to help with that interpretation. If you have used a process approach to teaching writing, the assessment in *Investigations* will seem familiar. For many of the assessment activities, a Teacher Note provides examples of student work and a commentary on what it indicates about student thinking.

Documentation of Student Growth

To form an overall picture of mathematical progress, it is important to document each student's work in journals, notebooks, or portfolios. The choice is largely a matter of personal preference; some teachers have students keep a notebook or folder for each unit, while others prefer one mathematics notebook, or a portfolio of selected work for the entire year. The final activity in each *Investigations* unit, called Choosing Student Work to Save, helps you and the students select representative samples for a record of their work.

This kind of regular documentation helps you synthesize information about each student as a mathematical learner. From different pieces of evidence, you can put together the big picture. This synthesis will be invaluable in thinking about where to go next with a particular child, deciding where more work is needed, or explaining to parents (or other teachers) how a child is doing.

If you use portfolios, you need to collect a good balance of work, yet avoid being swamped with an overwhelming amount of paper. Following are some tips for effective portfolios:

- Collect a representative sample of work, including some pieces that students themselves select for inclusion in the portfolio. There should be just a few pieces for each unit, showing different kinds of work—some assignments that involve writing, as well as some that do not.

- If students do not date their work, do so yourself so that you can reconstruct the order in which pieces were done.

- Include your reflections on the work. When you are looking back over the whole year, such comments are reminders of what seemed especially interesting about a particular piece; they can also be helpful to other teachers and to parents. Older students should be encouraged to write their own reflections about their work.

Assessment Overview

There are two places to turn for a preview of the assessment opportunities in each *Investigations* unit. The Assessment Resources column in the unit Overview Chart (pp. I-13–I-16) identifies the Teacher Checkpoints and Assessment activities embedded in each investigation, guidelines for observing the students that appear within classroom activities, and any Teacher Notes and Dialogue Boxes that explain what to look for and what types of student responses you might expect to see in your classroom. Additionally, the section About the Assessment in This Unit (p. I-20) gives you a detailed list of questions for each investigation, keyed to the mathematical emphases, to help you observe student growth.

Depending on your situation, you may want to provide additional assessment opportunities. Most of the investigations lend themselves to more frequent assessment, simply by having students do more writing and recording while they are working.

Mathematical Thinking at Grade 3

Content of This Unit   This unit introduces third graders to some of the content, processes, and materials they will be using to solve problems in mathematics. As the year begins, students do some exploring in each of the three areas of the *Investigations* curriculum: number (patterns, addition combinations, odds and evens, doubling and halving); data (collecting, organizing, representing, and describing information about themselves); and space (symmetrical designs with pattern blocks). This unit is meant to familiarize your students with *Investigations,* not to move them to mastery of particular mathematical concepts. Even if some of your students have only a partial understanding of some of the topics, don't spend too much time on them; your students will work with these concepts in more depth later in the year.

Throughout this unit, the students use basic mathematical tools and materials: interlocking cubes, 100 charts, calculators, money, and pattern blocks. They practice critical mathematical processes that they will be using all year: working with peers, writing and drawing about their work, and discussing alternative strategies for solving problems. This unit is also designed to help you assess students' strengths, needs, confidence, and flexibility in mathematics.

Connections with Other Units   If you are doing the full-year *Investigations* curriculum in the suggested sequence for grade 3, this is the first of ten units. It has connections with every other unit in the third grade sequence, both in its content and in its emphasis on ways of thinking and doing mathematics. There is a particular connection to the Addition and Subtraction unit, *Combining and Comparing,* where students will continue their work with strategies for addition and subtraction begun in this unit.

If your school is not using the full-year curriculum, this unit can be used at any time of the year as a way to help students focus on thinking, working, and talking mathematically, as well as to assess student understanding of some key mathematical content at this grade level. The unit can also be used successfully at grade 4.

Investigations Curriculum ■ Suggested Grade 3 Sequence

▶ *Mathematical Thinking at Grade 3* (Introduction)

Things That Come in Groups (Multiplication and Division)

Flips, Turns, and Area (2-D Geometry)

From Paces to Feet (Measuring and Data)

Landmarks in the Hundreds (The Number System)

Up and Down the Number Line (Changes)

Combining and Comparing (Addition and Subtraction)

Turtle Paths (2-D Geometry)

Fair Shares (Fractions)

Exploring Solids and Boxes (3-D Geometry)

Investigation 1 ▪ What's a Hundred?

Class Sessions	Activities	Pacing
Session 1 (p. 4) HOW MUCH IS 100?	Working in the Mathematical Environment How Much Is 100? Homework: 100 Objects	minimum 1 hr
Sessions 2 and 3 (p. 9) WORKING WITH 100	Teacher Checkpoint: Making a 100 Chart Introducing Plus–Minus–Stay the Same Two Choices: 100 Charts and Cubes Homework: Playing Plus–Minus–Stay the Same at Home	minimum 2 hr

◕ **Ten-Minute Math** ▪ **Calendar Math**

Mathematical Emphasis

- Counting and grouping quantities to make 100

- Becoming familiar with the number patterns on the 100 chart

- Exploring materials that will be used throughout this curriculum as problem-solving tools

- Communicating about mathematical thinking through written and spoken language

Assessment Resources

Observing the Students: How Much Is 100? (p. 6)

Double-Checking (Teacher Note, p. 8)

Teacher Checkpoint: Making a 100 Chart (p. 10)

Materials

Interlocking cubes

Colored pencils, markers, or crayons

Rubber bands or envelopes

Overhead projector and transparencies

Family letter

Student Sheets 1–4

Teaching resource sheets

Investigation 2 ■ Doubles and Halves

Class Sessions	Activities	Pacing
Session 1 (p. 16) PATTERN BLOCKS	Doubling with Pattern Blocks Symmetrical Pattern Block Designs	minimum 1 hr
Session 2 (p. 25) STRATEGIES FOR ADDITION	Discussion: Doubling Two-Digit Numbers Teacher Checkpoint: What Combinations Make 10? Which Are Hard? Which Are Easy? Learning Difficult Combinations Homework: Strategies for Addition	minimum 1 hr
Sessions 3 and 4 (p. 31) FINDING DOUBLES AND HALVES	Choice Time: Working with Doubles Homework: Problem Strategies Extension: Difficult Addition Combinations	minimum 2 hr
Sessions 5, 6, and 7 (p. 38) DOUBLING WITH MONEY	Teacher Checkpoint: Counting Money Choice Time: Doubles and Halves Discussion: Doing Doubles Problems Assessment: Writing about Addition Homework: Money Problems to Do at Home Homework: How Much Is Your Symmetrical Design Worth? Homework: More Addition in Two Ways	minimum 3 hr

◕ **Ten-Minute Math ■ Calendar Math**

Mathematical Emphasis	Assessment Resources	Materials
■ Constructing symmetrical patterns ■ Learning the addition combinations 1 + 1 to 10 + 10 ■ Developing and using addition strategies, including the use of known addition combinations to help learn others ■ Exploring what happens when 10 or 20 is added or subtracted ■ Exploring which numbers can be divided in half evenly ■ Reviewing coin values and finding the values of collections of coins	Strategies for Learning Addition Combinations up to 10 + 10 (Teacher Note, p. 22) Teacher Checkpoint: What Combinations Make 10? (p. 26) Two Powerful Addition Strategies (Teacher Note, p. 30) Teacher Checkpoint: Counting Money (p. 39) Assessment: Writing About Addition (p. 43)	Pattern blocks Mirror Chart paper Scissors Envelopes or plastic bags Interlocking cubes Coins Overhead projector and transparencies Student Sheets 5–12 Teaching resource sheets

Investigation 3 ▪ Data and Handfuls

Class Sessions	Activities	Pacing
Sessions 1 and 2 (p. 48) COLLECTING AND REPRESENTING DATA	Playing Guess My Rule Discussion: Can Data Change? Representing the Data from Guess My Rule Representing Data with Categories Homework: Data Analysis Homework: Calendar Math	minimum 2 hr
Sessions 3 and 4 (p. 59) HANDFULS OF CUBES AND OTHER OBJECTS	Handfuls of Cubes Grabbing Handfuls Looking at Data from Handfuls Handfuls: What Do You Think? Homework: Handfuls at Home Extensions: Handfuls from Other Students, Handfuls from Adults	minimum 2 hr

◗ **Ten-Minute Math** ▪ **Exploring Data**

Mathematical Emphasis	Assessment Resources	Materials

Mathematical Emphasis

- Sorting and classifying information

- Collecting, recording, and representing data

- Describing data presented in tallies and graphs

- Using grouping to count tallies or objects

- Developing strategies to combine and compare quantities

Assessment Resources

Observing the Students: Representing the Data from Guess My Rule (p. 52)

Sketching Data (Teacher Note, p. 57)

Using Concrete Materials for Addition and Subtraction (Teacher Note, p. 65)

Materials

Interlocking cubes

Paper

Chart paper

Colored pencils, markers, or crayons

Overhead projector and transparencies

Calculators

Containers of uniform items (such as beans, buttons, or pennies)

Student Sheets 13–17

Teaching resource sheets

Investigation 4 ▪ Exploring Odds and Evens

Class Sessions	Activities	Pacing
Session 1 (p. 70) ADDING ODDS AND EVENS	What Is an Even Number? Which Are Odd? Which Are Even? Extension: Subtraction with Odds and Evens	minimum 1 hr
Session 2 (p. 74) ODDS AND EVENS ON THE CALCULATOR	Teacher Checkpoint: Exploring the Calculator Discussion: What's the Little Dot? Splitting Numbers in Half Homework: Odd and Even Numbers	minimum 1 hr
Session 3 (p. 81) WHAT WE'VE LEARNED ABOUT ODDS AND EVENS	What We Found Out About Odd and Even Assessment: Writing About Odd and Even	minimum 1 hr

◔ **Ten-Minute Math** ▪ **Exploring Data**

Mathematical Emphasis

- Exploring the characteristics of odd and even numbers and examining how they behave when they are combined

- Using evidence gathered from examples to make conjectures about the ways numbers behave

- Continuing to develop familiarity with addition combinations

- Working with wholes and halves

- Developing awareness of the decimal point and its meaning

- Exploring mathematical tools such as the calculator

Assessment Resources

Teacher Checkpoint: Exploring the Calculator (p. 74)

Using the Calculator Sensibly in the Classroom (Teacher Note, p. 80)

Assessment: Writing About Odd and Even (p. 83)

Helping Students Clarify Their Ideas in Writing (Dialogue Box, p. 86)

Materials

Interlocking cubes

Calculators

Chart paper

Student Sheets 18–22

Teaching resource sheets

Following are the basic materials needed for the activities in this unit. Many of the items can be purchased from the publisher, either individually or in the Teacher Resource Package and the Student Materials Kit for grade 3. Detailed information is available on the *Investigations* order form. To obtain this form, call toll-free 1-800-872-1100 and ask for a Dale Seymour customer service representative.

Snap™ Cubes (interlocking cubes): at least 100 per pair

Pattern blocks: 1 bucket per 6–8 students

Plastic coins (real coins may be substituted): 50¢ to 70¢ per pair, in pennies, nickels, dimes, quarters

Numeral Cards (manufactured, or use blackline masters at the back of this book to make your own sets; optional)

Uniform items large enough to grab by handfuls and count (such as dried kidney or lima beans, color tiles, pennies, buttons)

Four or more open containers to hold the items to grab, such as shoe boxes or plastic tubs (with opening large enough for a student's hand)

Small rectangular mirror

Calculators: at least 1 per pair (ideally 1 per student)

Scissors: 1 per student

Colored pencils, markers, or crayons

Resealable plastic bags or envelopes (for storage)

Chart paper

Paper: plain, lined, and graph

The following materials are provided at the end of this unit as blackline masters. A Student Activity Booklet containing all student sheets and teacher resources needed for individual work is available.

Family Letter (p. 94)

Student Sheets 1–22 (p. 95)

Teaching Resources:

    How to Play Plus–Minus–Stay the Same (p. 99)

    Addition Cards (p. 108)

    Doubles and Halves Problems (p. 111)

    Choice List (p. 113)

    Money Problems (p. 114)

    One-Centimeter Graph Paper (p. 121)

    100 Chart (p. 127)

    Numeral Cards (p. 128)

Practice Pages (p. 131)

Related Children's Literature

Dee, Ruby. *Two Ways to Count to Ten*. New York: Henry Holt, 1988.

Hong, Lily Toy. *Two of Everything*. Morton Grove, Ill.: Albert Whitman, 1993.

O'Brien, Thomas C. *Odds and Evens*. New York: Thomas Y. Crowell, 1971.

Mathematical Thinking at Grade 3, as the title indicates, is designed to be an introduction to mathematical thinking—to some of the content, materials, processes, and ways of working that mathematics entails. The investigations in this unit engage students in:

- solving mathematical problems in ways that make sense to them
- talking, writing, and drawing about their work
- working with peers
- building models of mathematical situations
- relying on their own thinking and learning from the thinking of others

Students work with problems in the areas of number, data, and space (geometry). In number they look for patterns, explore properties of odd and even numbers, work on addition combinations, and begin to develop strategies for doubling and halving numbers and combining and comparing large quantities. Their work with geometry includes exploring mirror symmetry as they build designs with pattern blocks. In their study of data, students collect information about themselves as a group and begin to find ways of organizing, representing, and discussing that data.

Much of the work in this unit involves combining and comparing quantities. Students work with these ideas in a variety of contexts: building with pattern blocks and interlocking cubes, using money, dividing apples between two people, comparing the number of beans they can hold in their right and left hands, deciding if everyone in the class can have a partner, describing the kinds of shoes students in the class are wearing. They look at patterns on the 100 chart and discuss what they know about 100. They develop strategies for doubling quantities and for cutting quantities in half.

A major focus of these activities is the development and use of good number sense to combine and compare one-digit and two-digit numbers. Just as common sense grows from experience with the world and how it works, number sense grows from experience with how numbers work. Throughout these activities, students are encouraged to solve addition and subtraction problems by thinking about how the numbers are structured and how

they are related to other numbers. For example, in solving a one-digit addition combination they don't know, such as $8 + 9$, students are encouraged to use what they do know—perhaps $8 + 8$ or $10 + 8$— to reason about the sum. In solving two-digit addition or subtraction problems, we urge students not to apply rote procedures, but to look at the whole problem first and then apply what they know about the numbers to solve the problem.

The investigation of doubles and halves provides the foundation for students' exploration later in the unit of the characteristics of odd and even numbers. Looking at numbers and how they behave is not so different from observing, describing, and drawing conclusions about other kinds of data. When students collect and describe data about their class in Investigation 3, they think about similarities and differences within a group of people. In the final part of the unit, they examine similarities and differences of various numbers. What is it that makes 2, 4, 6, 8, 32, 100, and 126 the same? Why are they all "even"? How do these numbers behave? As students examine these numbers and model them with interlocking cubes, they begin to learn about and articulate important features of their structure. By gathering evidence and developing their own conjectures, they are entering the realm of number theory.

Mathematical Thinking at Grade 3 is designed not only to involve students with some central mathematical concepts, but also to introduce students to a particular way of approaching mathematics. Throughout the unit students are encouraged to share their strategies, work cooperatively, use materials, and communicate both verbally and in writing about how they are solving problems. These approaches may be quite difficult for some students. Even taking out, using, and putting away materials may be unfamiliar. Certainly writing and drawing pictures to describe mathematical thinking will be quite difficult for some students. This unit is a time to focus on the development of these processes; to spend time establishing routines and expectations; to communicate to students your own interest in and respect for their mathematical ideas; to assure students that you want to know about their *thinking*, not just their answers; and to insist that students work hard to solve problems in ways that make sense to them. As the unit unfolds, a mathematical community begins to take shape— a community that you and your students are together responsible for creating and maintaining.

Mathematical Emphasis At the beginning of each investigation, the Mathematical Emphasis section tells you what is most important for students to learn about during that investigation. Many of these mathematical understandings and processes are difficult and complex. Students gradually learn more and more about each idea over many years of schooling. Individual students will begin and end the unit with different levels of knowledge and skill, but all will gain greater knowledge about solving mathematical problems in number, data, and space (geometry), in ways that make sense to them.

Throughout the *Investigations* curriculum, there are many opportunities for ongoing daily assessment as you observe, listen to, and interact with students at work. In this unit, you will find four Teacher Checkpoints:

Investigation 1, Sessions 2–3:
Making a 100 Chart (p. 10)

Investigation 2, Session 2:
What Combinations Make 10? (p. 26)

Investigation 2, Sessions 5–7:
Counting Money (p. 39)

Investigation 4, Session 2:
Exploring the Calculator (p. 74)

This unit also has two embedded assessment activities:

Investigation 2, Sessions 5–7:
Writing About Addition (p. 43)

Investigation 4, Session 3:
Writing About Odd and Even (p. 83)

In addition, you can use almost any activity in this unit to assess your students' needs and strengths. Listed below are questions to help you focus your observation in each investigation. You may want to keep track of your observations for each student to help you plan your curriculum and monitor students' growth. Suggestions for documenting student growth can be found in the section About Assessment (p. I-10).

Investigation 1: What's a Hundred?

■ How do students count 100 objects? Do they count by 1's? 5's? 10's? Do they count in more than one way? How do they keep track of their count? How do they double-check? Do they use a different method to check?

■ What do students know about the structure of the numbers (for example, that the 30's come after the 20's, or that the 80's are closer to 100 than the 40's)? How do students use the 100 chart as a tool for exploring the number system? Do they know what happens when 10 is added to or subtracted from a two-digit number?

■ Do students actively explore the materials? After exploration time, do students use materials appropriately? Do they treat the materials with respect?

■ How do students verbally explain the strategies they use? Can they do this in more than one way? How do they represent this in writing? With words? With pictures? With both? Do they incorporate another's strategy into their own?

Investigation 2: Doubles and Halves

■ Do students' pattern block designs reflect mirror symmetry? How do students predict the total number of items in a symmetrical design after building only half of it?

■ What addition strategies do students naturally use? How do students figure out unfamiliar addition combinations? Do they use known combinations? Do they use the clues they devised? Do they have a strategy that can be applied to a wide range of unknown addition combinations (for example, adding all the 10's first or rounding up to familiar landmark numbers first)?

■ What strategies do students have for finding half of a number? Do they use their knowledge of addition combinations from 1 + 1 to 10 + 10? How do they find half of an odd number?

■ Do students know the values of coins? How do they add coins of like and unlike denominations?

■ How do students use materials (for example, cubes or 100 charts) appropriately to help them solve problems? Are they also using mental strategies based on their knowledge of numbers?

Investigation 3: Data and Handfuls

■ How do children sort and classify information? Do they name the category? Do they classify other types of items within that category? How do they test their theories? Do they use evidence? Examples? Counterexamples?

■ How accurately are students representing the data? Do they use a variety of ways (for example, concrete representations, line plots, pictures)? Do their sketches reveal important features of data? Are they clear? How do children change aspects of their representations that are not clear to others?

- How do students interpret or describe data?

- How are students organizing and keeping track of their data? Do they use groups (of tally marks or actual objects) when counting?

- What strategies do students use to find the difference between two numbers?

Investigation 4: Exploring Odds and Evens

- How do students describe an even number? An odd number? Do they use concrete examples? Do students know what happens when two odd numbers are added or subtracted? An odd number and an even number?

- Can students give reasons for their ideas? Do they use examples to help explain their ideas? Do they make generalizations from observed patterns and relationships?

- Do students know the solution to their addition combinations before they build and count them? How do they use the clues they've written on their cards?

- Can students split quantities in half? Are they comfortable splitting even numbers in half? What strategies do they use? With what range of numbers are they comfortable? Can they split odd numbers in half?

- Do students know, or are they beginning to be aware that 0.5 means one-half? Do they recognize the decimal point when they see it on the calculator screen and keyboard? What do they know about the meaning of the decimal point (for example, do they know that the decimal point always means that there is a part of the number that is less than one)?

- How comfortable are students with the calculator? Do they do straightforward computation easily? Are they familiar with the symbols on the keyboard? Do they read the screen? Do they recognize, interpret, and use symbols on the calculator keys to do addition, subtraction, multiplication, and division?

Thinking and Working in Mathematics

This unit provides the chance for you to observe students' work habits and communication skills. Think about these questions to help you decide which routines, processes, and materials will require the most ongoing support and guidance.

- Are students comfortable and focused working together in pairs? In small groups? How do they participate in discussions?

- Do students expect to devise their own strategies for solving problems, or do they expect you to tell them what to do? Do they understand that different people may solve problems in different ways?

- Are students familiar with the materials? Do they know how to use them? Do they use these materials as tools when solving problems? Do they take them out and put them away efficiently?

- Do students have ideas about how to record their work, or does writing and drawing about mathematics seem new to them?

- Can students choose an activity from among several, then move smoothly to a second activity when finished with the first?

In the *Investigations* curriculum, mathematical vocabulary is introduced naturally during the activities. We don't ask students to learn definitions of new terms; rather, they come to understand such words as *factor* or *area* or *symmetry* by hearing them used frequently in discussion as they investigate new concepts. This approach is compatible with current theories of second-language acquisition, which emphasize the use of new vocabulary in meaningful contexts while students are actively involved with objects, pictures, and physical movement.

Listed below are some key words used in this unit that will not be new to most English speakers at this age level but may be unfamiliar to students with limited English proficiency. You will want to spend additional time working on these words with your students who are learning English. If your students are working with a second-language teacher, you might enlist your colleague's aid in familiarizing students with these words before and during this unit. In the classroom, look for opportunities for students to hear and use these words. Activities you can use to present the words are given in the appendix, Vocabulary Support for Second-Language Learners (p. 91).

the numbers 1 to 100 Students use the 100 chart throughout the unit. They should be able to write the numerals and identify each by name.

money: coins, cents, penny, nickel, dime, quarter, dollar Students need to recognize U.S. coins and know the value of each as they work with money problems, doubling and halving, combining and comparing different amounts.

same, different, agree, disagree, check, double-check Students use the terms *same* and *different* as they compare sums and the combinations that make those sums. These terms are also an important part of checking and double-checking answers to problems.

add, plus, combine, more, less, most, least These terms are used throughout the unit as students combine quantities, discuss data they have collected, and use the calculator to find sums.

subtract, minus These terms are also used throughout the unit as students compare quantities and find the difference between them, with and without the calculator.

divide, equal, unequal, group As students explore halving and the characteristics of odd and even numbers, they discover what types of amounts will *divide* into *equal groups*.

pattern Students look for patterns on the 100 chart (for example, all the numbers in the last column end with zero); they also work with pattern blocks to make symmetrical designs.

Multicultural Extensions for All Students

Whenever possible, encourage students to share words, objects, customs, or any aspects of daily life from their cultures and backgrounds that are relevant to the activities in this unit. For example, when you are discussing U.S. coins and their values during Investigation 2, students might bring in coins from other countries. Suggest that they display them with their names and values.

Investigations

What's a Hundred?

What Happens

Session 1: How Much Is 100? Students count out 100 interlocking cubes and figure out ways to prove they have 100 cubes. They record their strategy using words, pictures, and numbers.

Sessions 2 and 3: Working with 100 Students enter numbers on a partially filled 100 chart, looking for patterns and discussing strategies for placing numbers. Then they use their 100 chart to play a number game and to construct something from the 100 blocks they counted out in Session 1.

Mathematical Emphasis

■ Counting and grouping quantities to make 100

■ Becoming familiar with number patterns on the 100 chart

■ Exploring materials that will be used throughout this curriculum as problem-solving tools

■ Communicating about mathematical thinking through written and spoken language

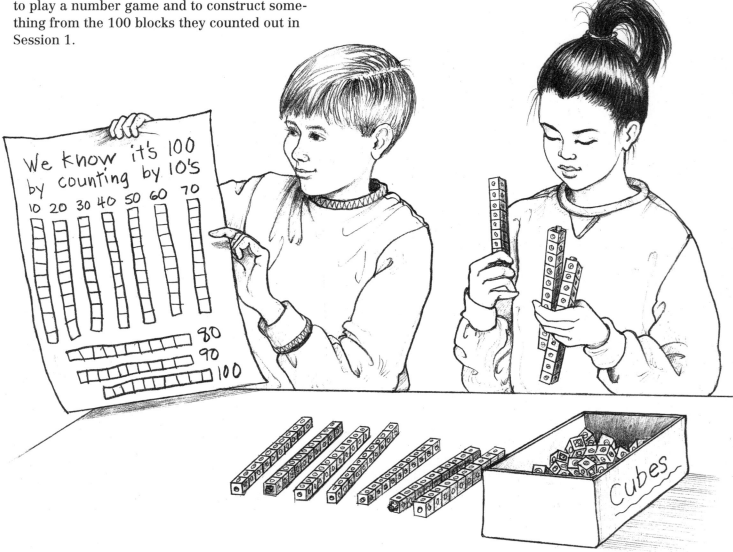

What to Plan Ahead of Time

Materials

- Interlocking cubes: at least 100 per pair (Sessions 1–3)
- Numeral Cards: 1 deck per pair (Sessions 2–3). If you do not have manufactured sets, make your own; see Other Preparation.
- Colored pencils, markers, or crayons (Sessions 2–3)
- Rubber bands or envelopes to hold cards: 1 per student, homework (Sessions 2–3)
- Overhead projector (Sessions 2–3)

Other Preparation

- Duplicate student sheets and teaching resources (located at the end of this unit) in the following quantities. If you have Student Activity Booklets, copy only the items marked with an asterisk.

For Session 1

Student Sheet 1, 100 Cubes (p. 95): 1 per student

Student Sheet 2, 100 Objects (p. 96): 1 per student (homework)

Family letter* (p. 94): 1 per student. Remember to sign the letter before copying.

For Sessions 2–3

Student Sheet 3, Partially Filled 100 Chart (p. 97): at least 1 per student, and 1 transparency*

Student Sheet 4, Numbers for the 100 Chart (p. 98): 1 per student

100 chart (p. 127): 5–6 per student (3–4 for homework), and 1 transparency*

Numeral Cards, pages 1–3 (p. 128): 1 each per student (homework)

How to Play Plus–Minus–Stay the Same (p. 99): 1 per student (homework)

- If you do not have manufactured Numeral Cards from the *Investigations* grade 3 materials kit, make 1 deck per pair for class use. If you duplicate the sheets on tagboard, the cards will last longer. Cut apart the 44 cards for one complete deck.
- Put interlocking cubes in buckets or shoe boxes for easy distribution to small groups of 4–6 students.
- Prepare a sheet of chart paper with the title *Patterns on the 100 Chart*. (Sessions 2–3)
- If you plan to provide folders in which students will save their work for the entire unit, prepare these for distribution during Session 1.

Materials

- Containers of interlocking cubes (at least 100 per pair)
- Student Sheet 1 (1 per student)
- Student Sheet 2 (1 per student, homework)
- Family letter (1 per student)

How Much Is 100?

What Happens

Students count out 100 interlocking cubes and figure out ways to prove they have 100 cubes. They record their strategy using words, pictures, and numbers. Their work focuses on:

- counting and grouping to 100
- exploring interlocking cubes
- using words, pictures, and numbers to record their thinking

Activity

Working in the Mathematical Environment

Begin this session with a brief discussion about some of the tools you'll be using and the kind of work students will be doing in this unit.

In our mathematics class this year, you will be using lots of different tools to help you solve problems and to show people how you are thinking about a problem. During the next few weeks, we will be using tools like calculators, pattern blocks, 100 charts, and interlocking cubes as we play math games and solve math problems.

❖ **Tip for the Linguistically Diverse Classroom** Ensure that this introduction is comprehensible to all students by showing the different kinds of mathematical tools as you mention them.

Ask students if they are familiar with any of these materials and how they have used them in the past.

Mathematicians use lots of different tools when they solve problems. Mathematicians also can show us how they think about and solve problems. They talk about their work, draw pictures, build models, and explain their work in writing so they can share their ideas with other people.

When you are working on a math problem, I will often ask you to use words, pictures, or numbers to explain how you solved it. Lots of times I will ask you to talk about how you solved a problem—either with a partner, in a small group, or with the whole class. This is one of the ways we can share good ideas for thinking about the math problems we are working on.

If you plan to use math journals or math folders, this is an appropriate time to introduce them. Explain that they are a way to collect the writing and drawing students will be doing during math class.

Exploring Interlocking Cubes   Introduce interlocking cubes as one of the tools students will be using during math class throughout the year. Allow 10–15 minutes for students to explore the cubes.

As with the introduction of any new material, it is important to establish clear ground rules. At the end of this exploratory time, ask students to check the floor for cubes and return all cubes to their containers (or push them to the middle of the table). Have a brief discussion about what students did with the cubes and what expectations you have for how they will be used and cared for. See the **Teacher Note**, Introducing Materials (p. 7), for hints about establishing routines for using and caring for manipulatives you'll be using.

One issue that may arise at this point is important to address: Often, you will want students to stop working and to focus their attention on a discussion. Because materials will likely be out on tables, some students may be distracted and continue playing with them during the discussion. It is important to set clear limits. Many teachers have found that giving a warning a few minutes before the end of a work period helps students be aware that a transition is about to occur. Establishing such routines is important throughout this introductory unit.

Though these first few sessions are designed to allow students to explore the cubes, you may want to provide other times during the day when students can build with the cubes and continue their exploration of this new material.

Finding 100   Pose the following task to the class:

Today, you and your partner are going to make a group of 100 cubes. When you think you have 100, find some way of organizing or arranging them so you can show us all that you have exactly 100. Then record, using words, pictures, and numbers, how you know you have 100 cubes. Tomorrow you and your partner will build something with your 100 cubes.

Working in pairs (or individually, if there are enough cubes), students count out 100 cubes from the containers. Students should be seated in small groups so that even if they are working individually, they will be able to discuss their strategies and ideas with other students. Encourage students to double-check their count by using more than one method of counting. See the **Teacher Note**, Double-Checking (p. 8), for more about this important process.

Observing the Students   As students work on this task, circulate around the room and observe the following:

- How are students counting? By 1's, 5's, 10's?
- How are students keeping track of their count?
- Do students have a way of double-checking their count?
- Are students accurate in their count?

Remind students that part of their task is to prove they have 100 cubes and to record their proof using words, pictures, and numbers. Distribute a copy of Student Sheet 1, 100 Cubes, to each student. Even if students have worked with a partner, they should each complete their own sheet.

Students will complete this task at various times. Encourage students who have completed the counting and recording to share their work with each other, explaining their counting strategies and how they can prove they have 100 cubes.

Third graders' organizational strategies will vary greatly. Some students will organize their cubes in groups of 10; others might make groups of 25 or 50. Many third graders will count their cubes one by one. Every approach will give you ideas about how your students are using counting and grouping strategies. By encouraging students to share their ideas with each other, either in small groups or in a class discussion, you will help them begin to see that there are many different ways of approaching the same task.

Students will need their collection of 100 cubes in Sessions 2 and 3. They can store them on pieces of paper (or in envelopes or on paper plates, if you have them) until the next session.

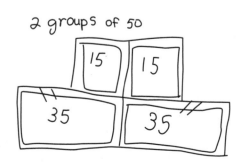

2 groups of 50

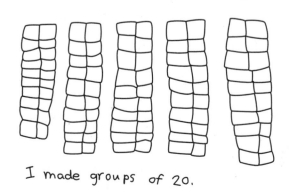

I made groups of 20.

100 Objects   After Session 1, students can practice grouping by counting out 100 objects at home. Suggest that students use objects such as pennies, buttons, toothpicks, Cheerios, or macaroni to count at home. They can use Student Sheet 2, 100 Objects, to record their strategies for counting out 100 objects. Also send home the family letter or *Investigations* at Home.

 **Homework**

Introducing Materials

Teacher Note

Using concrete materials in the classroom may be a new experience for many students and teachers. Before introducing new materials, think about how you want students to use and care for them and how they will be stored.

Introducing a New Material   Students will need time to explore a new material before using it in a structured activity. By freely exploring a material, students will discover many of its important characteristics. Once having had the chance to make their own decisions about how they would like to use a material, students will be more ready to use it as a tool in a specific activity. Although some free exploration should be done during regular math time, many teachers make materials available to students during free time or before or after school.

Establishing Routines for Using Materials
Establish clear expectations about how materials will be used and cared for. Students generally find math manipulatives attractive and engaging, and are eager to use them. Consider having students suggest rules for how materials should and should not be used. Students will often be more aware of rules and policies that they have helped create. You might want to establish the rule that if students are unable to use materials responsibly, they lose the opportunity to use them for a period of time.

Plan for how materials will be distributed and cleaned up at the beginning and end of each class. Some teachers assign one or two students each week to be responsible for passing out and collecting materials. Most teachers find that stopping 5 minutes before the end of class gives students time to clean up materials, put their work in their folders, and double-check the floor for any stray materials.

Caring For and Storing Materials   Store manipulatives where they are easily accessible to students. Many teachers store materials in plastic tubs or shoe boxes arranged on a bookshelf or along a windowsill. Students should view materials as useful tools for solving problems or illustrating their thinking—the more available they are and the more frequently they are used, the more likely students are to use them! Model this process by using manipulatives yourself when you are solving a problem, and encourage all students to use them. If manipulatives are only used when someone is having difficulty, students can get the mistaken idea that using materials is a less-sophisticated way to solve a problem.

Mathematicians frequently use concrete materials and build models to solve problems and to explain their thinking. Encourage your students to think and work like mathematicians!

A prevailing myth about mathematics is that if you are good at it, you always get the right answer the first time. On the contrary, for many mathematicians and others who are competent in using numbers, the key to good work is not initial precision and accuracy, but their use of multiple strategies to check and double-check their own problem-solving processes.

When counting a group of objects—Do we have enough chairs? How many more nails do I need? How many people are here?—almost anyone can miscount. To count accurately, we organize our counting by putting things into groups such as 2, 5, or 10. We also double-check our count, either by counting more than once or by having several people count.

Similarly, when solving a computation problem, we should check one strategy with another. Doing a problem slowly and carefully may not be enough. It is too easy to make a mistake—and then to make the same mistake again. A better way of double-checking is to use two different methods. Add the numbers in your head, then add them on the calculator. Add them from right to left, then from left to right or from bottom to top, or group them in a different way.

One way of double-checking is by learning strategies from each other: I did it this way; show me how you did it so we can double-check my solution. You don't need to teach specific double-checking strategies, but can instead point out various ways of solving a problem by having students share ideas. Students must double-check in ways that make sense to them and that are built firmly on their own number sense.

Learning double-checking strategies is a key part of becoming a confident mathematics user. Emphasize throughout the year that getting an accurate solution right away is not what counts. What's important is knowing that you have double-checked in such a way that you can be confident of the accuracy of your solution.

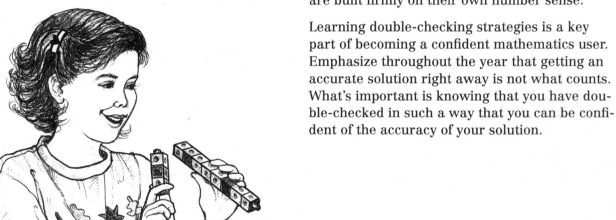

Working with 100

What Happens

Students enter numbers on a partially filled 100 chart, looking for patterns and discussing strategies for placing numbers. Then they use their 100 chart to play a number game and to construct something from the 100 blocks they counted out in Session 1. Their work focuses on:

- exploring the 100 chart
- looking for and using number patterns on the 100 chart
- counting by 10's on the 100 chart
- building with 100 cubes

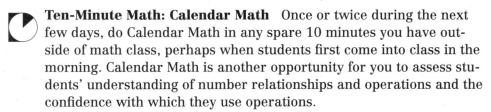

Ten-Minute Math: Calendar Math Once or twice during the next few days, do Calendar Math in any spare 10 minutes you have outside of math class, perhaps when students first come into class in the morning. Calendar Math is another opportunity for you to assess students' understanding of number relationships and operations and the confidence with which they use operations.

Write today's date on the board or ask students to find it on the calendar. Then, ask students to give number combinations that equal this number. List responses on the board. For example, if the date is September 10, solutions could include $5 + 5$, $12 - 2$, $100 - 90$, and $5 + 1 + 1 + 1 + 1 + 1$.

Have the class choose a "favorite expression" for the day from the list. Write the date on the board using this expression; for example, September $10 - 0 + 0$.

For variations on this activity, see p. 87.

Materials

- Transparency of Student Sheet 3 (optional)
- Student Sheet 3 (1 per student)
- Student Sheet 4 (1 per student)
- Prepared chart, Patterns on the 100 Chart
- Transparency of the 100 chart
- Numeral Cards (1 deck per pair)
- 100 chart (2 per student, plus 3–4 extras for homework)
- Colored pencils, markers, or crayons
- Students' collections of 100 cubes from Session 1
- Numeral Card sheets (1 set per student, homework), with rubber bands or envelopes to hold cards
- How to Play Plus–Minus–Stay the Same (1 per student, homework)
- Overhead projector

Teacher Checkpoint

Making a 100 Chart

Teacher Checkpoints are places for you to stop and observe student work (for more information, see About Assessment, p. I-10). However, keep in mind that this entire unit is designed to help you assess your students' understanding of mathematical ideas. See About the Assessment in This Unit (p. I-20) for detailed questions you can ask yourself about student understanding during each investigation.

Put the transparency of Student Sheet 3 on the overhead projector. (Or, draw the incomplete 100 chart on the board. You could use a pocket 100 chart, with selected numbers covered, if you have one.)

Write a number, such as 23, next to the incomplete 100 chart. Ask students where the number belongs on the chart. Solicit ideas from more than one student, and encourage students to explain their thinking.

Write another number, this time from a different row and column, and ask students where it belongs. Do this for three or four numbers, each time encouraging students to explain their thinking about the placement of the number.

Seat students in small groups (3–5 per group). Explain that they will be making their own number chart that will include numbers from 1 to 100. Distribute to each student a copy of Student Sheet 3, Partially Filled 100 Chart, and Student Sheet 4, Numbers for the 100 Chart.

As students begin to fill in their charts with the given numbers, encourage them to look for patterns and clues that help them place each number, and to talk with each other about their work. If they want, they may work with a partner.

When students have placed all the listed numbers, they fill in the remainder of the 100 chart. Encourage them to find ways to double-check their work; see the **Teacher Note**, Double-Checking (p. 8). One way is to compare their charts with one another. An important benefit of seating students in groups is that they can use each other as resources. You may need to encourage and even give them permission to discuss their work and to ask each other questions when they do not understand or agree.

Observing the Students As students place numbers on their charts, the following observations can provide you with information about their familiarity with numbers and number patterns:

- Do students recognize the numbers on their list?
- How are they placing numbers on the chart?

- Do they know that a number in the 20's will be near the top of the chart and a number in the 80's will be near the bottom?
- Are they counting from 1 for every number they add?
- Are they looking for patterns? (For example, "All the numbers in the 20's go in this row.")
- Do they skip count by 10's? ("I know 53 would go in the 3's column: 3, 13, 23, 33, 43, 53.")
- How do they fill in the remainder of their chart? Do they count from 1 or do they use number patterns?

Discussing the 100 Chart   When most students have completed their charts, have a brief discussion about patterns they noticed. Put a transparency of a completed 100 chart on the overhead so you and the students can refer to specific numbers. List the patterns students identify on your prepared chart, Patterns on the 100 Chart. Post this chart in the classroom so you can refer and add to it throughout this unit. See the **Dialogue Box**, Patterns on the 100 Chart (p. 13), for examples of patterns third graders have noticed.

While discussing patterns on the 100 chart, students will probably notice and be able to self-correct mistakes and misplaced numbers in their own work. Once again, remind them to double-check their work. Some students may want to make a new 100 chart if their original is difficult to read or too hard to correct. Have extra copies of Student Sheet 3 available.

Introducing Plus–Minus– Stay the Same

Introduce the game Plus–Minus–Stay the Same. You will be sending home the complete directions (p. 99) with the homework, but these sheets should not be needed in class.

The 100 chart is one of the tools we will be using a lot during math class this year. I want to teach you a game that can be played on the 100 chart. It's called Plus–Minus–Stay the Same. You play this game with a partner. The object is to cover five numbers in a row on your chart before your partner does.

Demonstrate by playing Plus–Minus–Stay the Same on the overhead. Choose two Numeral Cards from the deck to get your base number. Explain that the first number will tell you how many tens, and the second number will tell how many ones. For example, if you choose 3 and then 6, your number is 36. If you choose a 0 and a 7, your number is 07, or 7.

Explain that you must now decide what number to cover on your 100 chart. You are allowed to add 10 to the number, subtract 10 from it, or stay with the same number. Write the three possible numbers on the board and talk through your decision about which number you choose. Use a colored pen to color the number on the chart.

Demonstrate this procedure once or twice, then ask a student to choose two Numeral Cards, to determine the base number, and to suggest which numbers they could cover by adding 10 to the base number, subtracting 10 from it, or staying the same. Tell students that when they play this game, each player will have a separate 100 chart. Two players will take turns picking Numeral Cards and covering a number on their chart.

It is not necessary to complete an entire game on the overhead. When several numbers have been covered, remind students that the object is to cover five numbers in a row. Ask about possible strategies they might use to determine which number is the best choice.

Two Choices: 100 Charts and Cubes

When you feel students have grasped the rules of Plus–Minus–Stay the Same, give them the following choices:

For the remainder of today's session and tomorrow during math, you will need to complete two activities. First, you must play at least one game of Plus–Minus–Stay the Same.

Second, you must build something with the 100 cubes you and your partner counted out [yesterday]. Then you will write a short description and draw a picture of what you built. Use plain paper for this. You can start with either activity, but you must complete both by the end of tomorrow's class.

For Plus–Minus–Stay the Same, each student will need a 100 chart for each game (plan on at least two games), plus a crayon or marker for coloring in the number squares, and each pair will need one deck of Numeral Cards. For the cube-building activity, provide paper and colored pencils, markers, or crayons for student pictures and descriptions.

As students work on these choices, you may want to meet with pairs or small groups that either need help or that you haven't had a chance to observe before.

You may want to arrange a space in the classroom for students to display their 100-cube projects. After students have had a chance to view each other's buildings, have them return the cubes to their containers in towers of ten.

 **Homework**

Playing Plus–Minus–Stay the Same at Home   After Session 3, students can play Plus–Minus–Stay the Same at home. They will need a few copies of the 100 chart, sheets 1–3 of Numeral Cards to cut up and make into a deck, a rubber band or envelope to hold their Numeral Cards, and the How to Play Plus–Minus–Stay the Same directions. If students do not have scissors at home, suggest they cut up their home deck of Numeral Cards at school.

Ask students to teach the game to someone in their family. Suggest that they use pennies, buttons, or cut-up bits of paper to cover the numbers so that the 100 charts can be reused.

D I A L O G U E B O X

Patterns on the 100 Chart

In this discussion during the activity Making a 100 Chart (pp. 10–11), students describe patterns they discover on the 100 chart.

Tamara: All the counting-by-10's numbers go straight down the side of the chart. See: 10, 20, 30, 40, …

Sean: Yeah, and that's the same for other numbers, too. Like all the numbers with one at the end, 21, 31, 41, 51—they all go straight down. That's how I filled in my chart, by going down.

Tyrell: I see another pattern about the counting-by-10's numbers. If you go down each row, all the 1's are in the top row, then the 10's, then the 20's, then the 30's, all the way to the 90's.

Lots of you are noticing that there are patterns going across the rows and down the columns that have to do with counting by ten.

Chantelle: I see a different pattern. If you start at the 1 corner and go down to the 100 corner …

Dylan: That's the diagonal.

Chantelle: Yeah, if you go on the diagonal from 1 to 100, then the 1's number goes up by 1 and the 10's number goes up by 1. See: 12, 23, 34, 45.

Dylan: There's another one that does that: 13, 24, 35, 46, 57, 68, 79.

Chantelle: Hey—that's true for all the diagonals that go from left to right!

Dylan: And look, when the diagonal goes the other way [*he gestures right to left*] the numbers go up by 10's but down by 1.

Doubles and Halves

What Happens

Session 1: Pattern Blocks Students make a pattern block design with line, or mirror, symmetry. They build half the design on one side of the mirror line, count the number of blocks in that half, then predict how many blocks they will need for the total design. The class makes a list of the doubles—the addition combinations from 1 + 1 to 10 + 10—to use in solving problems that involve doubling.

Session 2: Strategies for Addition Students discuss strategies for doubling two-digit numbers, using their work with pattern blocks as a context. The class makes a list of all the addition combinations with a sum of 10. Then students make a set of Addition Cards, which they will use as they develop strategies for learning the addition combinations to 10 + 10. Students sort their cards into two sets—the combinations they know, and the combinations they need to learn. The class discusses possible strategies for learning difficult combinations.

Sessions 3 and 4: Finding Doubles and Halves Students are introduced to Choice Time, during which they choose from among several activities: working with symmetrical pattern block designs, playing Plus–Minus–Stay the Same, working out problems about doubles and halves, and learning some of the addition combinations they found difficult earlier in the investigation.

Sessions 5, 6, and 7: Doubling with Money Choice Time continues after students participate in an introductory activity with coins, which is also a Checkpoint for you to look at their familiarity with coin values. A new choice, Money Problems, is added to Choice Time. Students discuss and compare strategies for doubling problems. In an assessment activity, students solve a two-digit addition problem in two ways and record their work.

Mathematical Emphasis

- Constructing symmetrical patterns
- Learning the addition combinations 1 + 1 to 10 + 10
- Using known addition combinations to help with learning others
- Developing and using addition strategies
- Exploring what happens when 10 or 20 is added or subtracted
- Exploring which numbers can be divided in half evenly
- Reviewing coin values and finding the values of collections of coins

What to Plan Ahead of Time

Materials

- Pattern blocks: 1 bucket per 6–8 students (Sessions 1, 3–7)
- Small rectangular mirror (Session 1)
- Chart paper (Sessions 1–2)
- Scissors: 1 per student (Session 2)
- Envelopes or resealable plastic bags for storing Addition Cards: 1 per student (Session 2)
- Interlocking cubes: 100 per pair (Sessions 2–7)
- Numeral Cards (Sessions 3–7)
- Plastic coins: a few pennies, nickels, dimes, and one quarter per pair, in differing total amounts from 50¢ to 70¢ (real coins may be substituted) (Sessions 5–7)
- Overhead projector (Sessions 5–7)

Other Preparation

- Duplicate student sheets and teaching resources (located at the end of this unit) in the following quantities. If you have Student Activity Booklets, copy only the items marked with an asterisk, including any extra materials needed.

For Session 1

Student Sheet 5, Half and Half (p. 100): 1 per pair, plus extra copies for Choice Time in Sessions 3–7, and 1 transparency*

Student Sheet 6, How Many Blocks? (p. 101): 1 per student, plus 1–2 per student for Choice Time in Sessions 3–7

For Session 2

Addition Cards (p. 108): 1 each per student (plus 1 set for homework, if you don't want to send home the class sets).

Student Sheet 7, Strategies for Addition (p. 102): 1 per student (homework)

For Sessions 3–4

Student Sheet 8, Problem Strategies (p. 103): 3–5 per student (2 for Choice Time in Sessions 5–7, 1 for homework)

Doubles and Halves Problems (p. 111): 5–6 copies, cut apart and placed in envelopes, or 1 copy each per student, cut apart and assembled as small booklets

100 chart (p. 127): 4–5 per student (for use throughout Choice Time)

Choice List (p. 113): 1 per student (optional)

For Sessions 5–7

Student Sheet 9, Money Problems to Do at Home (p. 104): 1 per student (homework)

Student Sheet 10, How Much Is Your Symmetrical Design Worth? (p. 105): 1 per student (homework)

Student Sheet 5, Half and Half (p. 100): 1 per student (homework)

Student Sheet 11, Addition in Two Ways (p. 106): 1 per student

Student Sheet 12, More Addition in Two Ways (p. 107): 1 per student (homework), plus extras for modifying the assessment activity*

Money Problems (p. 114): 5–6 copies, cut apart and placed in boxes or envelope*

Pattern Blocks

Materials

- Pattern blocks (1 bucket per 6–8 students)
- Small rectangular mirror
- Transparency of Student Sheet 5
- Student Sheet 5 (1 per student)
- Student Sheet 6 (1 per student)
- Chart paper

What Happens

Students make a pattern block design with line, or mirror, symmetry. They build half the design on one side of the mirror line, count the number of blocks in that half, then predict how many blocks they will need for the total design. The class makes a list of the doubles, the addition combinations from 1 + 1 to 10 + 10, to use in solving doubling problems. Students' work focuses on:

- exploring with pattern blocks
- constructing a design with mirror symmetry
- developing strategies to add two numbers
- finding numerical patterns
- using known addition combinations to figure out others

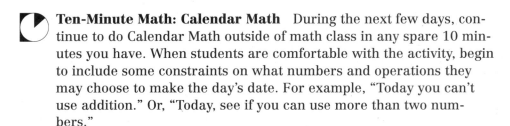 **Ten-Minute Math: Calendar Math**   During the next few days, continue to do Calendar Math outside of math class in any spare 10 minutes you have. When students are comfortable with the activity, begin to include some constraints on what numbers and operations they may choose to make the day's date. For example, "Today you can't use addition." Or, "Today, see if you can use more than two numbers."

As students work with these constraints, assess their comfort with a variety of numbers and operations. For full directions and variations on this activity, see p. 87.

Doubling with Pattern Blocks

Give each student a copy of Student Sheet 5, Half and Half. Use the transparency of this sheet or draw a model on the board as you introduce the activity.

I'm giving you a piece of paper with a line across the center. Turn your paper this way. [*Demonstrate by orienting the paper with the long side horizontal and the line vertical.*] **We're going to call this line your mirror line. That is to say, if this line were a *mirror* and I put two blocks on one side of the line, what would the *reflection* of the blocks look like?**

❖ **Tip for the Linguistically Diverse Classroom** Have a small mirror handy to help convey the ideas *mirror-image* and *reflection.*

Illustrate by drawing a very simple example on the board, or by using pattern blocks on the overhead.

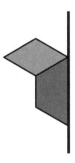

How many blocks do I have on this side of my design? How many blocks will I have in the whole design when I make the other half?

Have students place the same two shapes on their copy of Half and Half, and then show what they think would happen on the "mirror side" of the paper. Observe the students. If there are differences of opinion, show differing ideas on the board or overhead and ask students to discuss them. Students can check their designs by holding your mirror on the mirror line of the student sheet to see what the mirror-image of the design would look like.

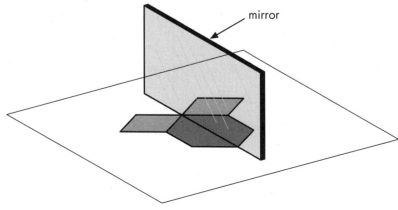

mirror

Tell students that patterns that reflect over a line have *mirror symmetry* or *line symmetry.* Continue using these words to describe students' patterns, but don't insist that they use them. See the **Teacher Note,** Using Mathematical Vocabulary (p. 21).

Try another example or two until most students seem comfortable with the idea of a symmetrical design. Notice which students are likely to need your help or to need the mirror to make their own designs.

Looking at Doubles   Begin a list of addition combinations of doubles up to 9 + 9 by asking:

Suppose I had 3 pattern blocks on one side of my design. How many would I have in the whole design? How do you know? If I had 4 blocks on one side, how many would there be in the whole design? How do you know?

Write the heading *Doubles* on a piece of chart paper. Record student answers on the chart, leaving room to continue the list in both directions:

$$3 + 3 = 6$$
$$4 + 4 = 8$$

Many students will probably say they "just know" these doubles (we refer to these addition combinations as "the doubles" throughout this unit). Students often learn the doubles before other single-digit addition combinations. Encourage students to contribute other doubles as you fill in the list.

What are some other doubles that you know? Let's see, no one mentioned [9 + 9]. Does anyone know what that is? How did you figure it out?

Encourage students to explain how they know the totals for some of the more difficult doubles. Some students may continue to say "I just know," but then you can ask, "Does anyone have a good way to figure it out if someone doesn't just know it?"

Encourage and model good mental arithmetic strategies, such as using a known fact to solve an unknown one: "7 + 7 is like 6 + 6, with 1 extra on each number. I know that 6 + 6 is 12, so I add two more, 13, 14." For examples of similar strategies, see the **Teacher Note,** Strategies for Learning Addition Combinations up to 10 + 10 (p. 22). If none of your students volunteer strategies like these, you can introduce them:

Here's a method some people use. Does it work? Can you use it to solve the next one?

Students may want to add 0 + 0 and 10 + 10 to the list. Once your list is complete, ask students to describe anything they notice about the list.

I put the doubles in order because mathematicians often put things in order to look for patterns. Do you see any patterns in our list?

Students may offer such observations as, "When the numbers go up by 1, the total goes up by 2," or, "They all add up to even numbers." This discussion relates to the work students will be doing on odd and even numbers in Investigation 4. You may want to pose a couple of questions for them to think about:

If the list kept going, would it always go up by 2? Why? Would the sums always be even?

Don't make students come to conclusions now. Have them contribute some initial ideas, then tell them you will return to the discussion in a few days. Leave the list of doubles posted so students can use it as a reference.

Symmetrical Pattern Block Designs

Distribute Student Sheet 6, How Many Blocks?, to each student. During the rest of the session, students work in pairs to make designs with pattern blocks on Student Sheet 5, Half and Half, and use their designs to explore halves and doubles. Each student records his or her findings individually on Student Sheet 6, which has room for exploring two designs. If students want to do more designs, suggest that they continue to record their findings in their mathematics notebooks.

Provide these instructions, possibly writing them on the board for reference:

1. Make a design on one side of the line.
2. Count how many blocks are in that half of the design.
3. Figure out how many blocks will be in the whole design. Record your prediction and how you figured out the total. (Use your sheet How Many Blocks? or your mathematics notebook.)
4. Build the rest of the design.
5. Count the pieces in the whole design and record this number.

❖ **Tip for the Linguistically Diverse Classroom** Pair students who may have difficulty reading Student Sheet 6 with English-proficient students who can read the directions aloud, using gestures and drawings as necessary to clarify meaning. Students who have difficulty writing in English may use drawings to explain their predictions.

As students work on their designs, walk around the room and observe. Insist that students double-check their counts by using more than one counting method and by having more than one person count. Ask:

How do you know your count is accurate? How do you keep track of which blocks have already been counted?

When students make their predictions about how many blocks will be in the whole design, make sure they describe how they figured out the total. Part of the purpose of this activity is to help students get used to the expectation that they will describe their strategies, not just record answers, and also to help them understand that you are interested in *how they think,* not just in their answers. For an example of a teacher helping a student record, see the **Dialogue Box,** Recording Strategies (p. 24).

When it is time to put the pattern blocks away, emphasize the procedure for storing materials: returning them to their containers, checking the floor for strays, and storing the containers in the designated area.

Note: Students may want to preserve their pattern block designs. Recording designs is a hard task, but tracing around the blocks, or using pattern-block templates or stickers (if you have them) can help. If you have enough pattern blocks, students can display their designs on a shelf or table (at least until the next session, when you will need the pattern blocks again). Or, you might have students walk around and view each other's designs before they put away the pattern blocks.

Using Mathematical Vocabulary

Students learn mathematical words the same way they learn other vocabulary—by hearing them used correctly, frequently, and in context. Young children learn words by hearing their families use language appropriately. When they make mistakes—"Look at the big doggie," says a young child pointing at a horse—adults use the correct words, and the child gradually learns the distinctions among all the four-footed animals.

We don't ask young children to memorize definitions of *horse* or *dog;* rather, we find opportunities for them to hear words being used in meaningful ways. Learning mathematical vocabulary is no different. Students do not learn how to speak mathematics by memorizing definitions, but by hearing these words frequently and having many opportunities to use them in context.

Use mathematical vocabulary accurately and frequently, and connect it with more familiar words that students may know:

What can you say about the number of people wearing shoes with laces *compared* to the number of people wearing sandals? You think there are 5 *more* people wearing shoes with laces? How did you figure out that the *difference* is 5?

Let's look at Ly Dinh's pattern block design. Does it have *line (mirror) symmetry?* Is one side a *reflection* of the other? How can you tell? Is there anything in the classroom that is *symmetrical?*

Throughout the *Investigations* curriculum, we will point out mathematical terms that are important for you to use in context. Don't insist that the students use these terms. What's important is that they express their ideas and describe their strategies for solving mathematical problems clearly and accurately, using whatever words are comfortable for them.

As you use mathematical terms frequently in context, students will become used to hearing them and will begin to use them naturally. Even young children can learn to use mathematical vocabulary accurately when they hear it used correctly and in the context of meaningful activities.

To develop good computation strategies, students must become fluent with the addition combinations from 0 + 0 to 10 + 10. These combinations are part of the repertoire of number knowledge that contributes to the rich interconnections among numbers we call *number sense*. A great deal of stress has been put on learning these addition combinations in elementary school. While we agree that knowing these combinations is important, we want to stress two important ideas:

■ Students learn these combinations best by using strategies, not by memorization. If you forget—as we all do at times—you are left with nothing. If, on the other hand, you've learned by using strategies based on knowledge of numbers and their relationships, you have a way to rethink and recall when you don't remember something you thought you "knew."

■ Knowing the facts is judged by fluency in use, not by instantaneous recall. In this curriculum, students are not expected to solve a certain number of addition combinations in a certain amount of time. Rather, through repeated use and familiarity, they come to know most of the addition combinations immediately, and a few by using quick and comfortable numerical reasoning strategies.

This is an approach that serves even adults. For example, when one of the *Investigations* authors thinks of 8 + 5, she doesn't automatically see the total as 13; rather, she sees the 5 broken apart into a 2 and a 3, the 2 combined with the 8 to make 10, and the 10 and the 3 combined to total 13. While this strategy takes a while to write down or to read, she sees this relationship almost instantaneously. As far as she is concerned, she knows this addition combination.

Strategies Students Can Use Many of the 121 addition "facts" from 0 + 0 to 10 + 10 are learned without difficulty. As our starting point in this

unit, we assume students know the combinations that involve + 0, + 1, and + 2. Usually they solve these combinations quickly by counting on. However, they also need to recognize that, for example, adding 2 + 8 is the same as adding 8 + 2, so that they can use the more efficient counting-on strategy (8, 9, 10) rather than beginning with 2 and counting up 8 more.

If you have students who have trouble with addition using zero, they might make Addition Cards for these combinations to add to their set, or model these combinations using cubes.

Once the + 0, + 1, and + 2 combinations are known, 36 combinations up to 10 + 10 remain if pairs of combinations (such as 7 + 3 and 3 + 7) are learned together. These appear on the Addition Cards (pp. 108–110), and they can be grouped to help students learn good strategies for solving them easily. Here are some useful groupings:

■ **The Doubles**—from 3 + 3 to 10 + 10. Students learn most of the doubles readily and can use the doubles they know to help with the harder doubles: "I know that 6 + 6 is 12, so 7 + 7 is 2 more: that's 14."

■ **The 10 + Combinations**—from 10 + 3 to 10 + 10. Because these facts follow a structural pattern, students learn these readily once they have built them repeatedly with cubes or counted them out on the 100 chart.

■ **The Near Doubles**—3 + 4, 4 + 5, 5 + 6, 6 + 7, 7 + 8, 8 + 9, and 9 + 10. These are 1 away from the doubles. Students can use the doubles they know to learn these: "I know that 5 + 5 is 10, so 5 + 6 is one more," or, "I know that 6 + 6 is 12, so 5 + 6 is one less."

■ **The 9 + Combinations**—from 9 + 3 to 9 + 10. Students can think of these this way: "To solve 9 + 6, I take one from the 6, add it to the 9 to make it 10, then have 5 left: 10 + 5 is 15," or, "If this were 10 + 6, the answer would be 16, but it's one less, so it's 15."

■ **Sums That Make 10**—3 + 7, 4 + 6, 5 + 5, 6 + 4, and 7 + 3. Students need many experiences building all the ways there are to make 10 with interlocking cubes until they recognize these combinations.

Some combinations belong to more than one group: 9 + 8 is a 9 + combination and also a near double. Students can use whatever way works best for them.

Only eight single-digit addition combinations do not fall into any of the categories above: 5 + 3, 6 + 3, 7 + 4, 7 + 5, 8 + 3, 8 + 4, 8 + 5, and 8 + 6. Many strategies can be used for each of these facts: relating them to doubles, to combinations that make 10, or to the 10 + combinations. Students should choose whatever clues make the most sense to them. Here are some possibilities:

Clues for 8 + 5

■ Take 2 from the 5, add it to the 8 to make 10; there's 3 left: 10 + 3 = 13. (Student may write 8 + 2 or perhaps 10 + 3 as a clue.)

■ 5 + 5 is 10, then 3 more from the 8 makes 13. (Student writes 8 + 2 + 3, or just 8 + 2, as a clue.)

■ I know that 10 + 5 is 15, so 9 + 5 is 14, and 8 + 5 is 13. (Student writes 10 + 5 as a clue.)

Clues for 7 + 5

■ Take one from the 7 and add it to the 5 to give you 6 + 6, which is 12. (Student writes 6 + 6.)

■ 5 + 5 is 10, then 2 more from the 7 makes 12. (Student writes 5 + 5.)

■ 7 + 3 is 10, and 2 more from the 5 makes 12. (Student writes 7 + 3.)

Note: You will notice that we have avoided calling these addition combinations "facts." We prefer to avoid the term *facts* because it tends to elevate knowledge of these combinations above other mathematical knowledge—as if knowing these is the important thing in mathematics. Developing fluency with these combinations is important, but many other ideas are just as critical for developing sound number sense.

8 + 5
5 + 8
Clue: 8 + 2 = 10

8 + 5
5 + 8
Clue: 10 + 5

7 + 5
5 + 7
Clue: 6 + 6

7 + 5
5 + 7
7 + 3 + 2
Clue:

Recording Strategies

These students are working on the activity Symmetrical Pattern Block Designs (p. 19). Like other students who are not accustomed to writing about their reasoning in mathematics class, these students had some difficulty writing down their reasoning on Student Sheet 6, How Many Blocks?

Ryan made a pattern block design with 16 blocks in one half. He wrote, "I think there will be 32 blocks in the whole design because 16 and 16 is 32." Looking at what he wrote, the teacher asked:

But why do you think 16 and 16 add to 32? How did you figure that out?

Ryan [*with a shrug*]: I just counted.

How did you count? Show me how you did it.

Ryan: Well, I counted this half, but I counted two for every block. See: 2, 4, 6, 8, 10—like that.

Why did you think that would tell you how many blocks there would be in the whole design?

Ryan: Because for every one here, there's one here, so it's like you're counting two at once.

You need to write more about how you did this so someone who didn't see you do it would understand.

Ryan: I don't get what to write.

Write what you told me—that you counted by 2's—and about why you did that. Add something about that, and then come show me what you wrote.

Latisha had 17 blocks in half of her design. She wrote, "I think there will be 34 blocks in the whole design because I added them." Again, the teacher asked her to explain more about what she did.

How did you add?

Latisha: I put 17 and 17, so there's 10 and 10, that would be 20. Then 7 and 7 is, um, 14. So it's 34.

I think I understand. You had 20 and 14. How did you know that was 34?

Latisha: Because 20 and 10 more is 30, then 4 more.

You figured out a very good strategy to add 17 and 17. You need to write that part down so I can remember what you did. You can use words or numbers to explain what you did.

Strategies for Addition

What Happens

Students discuss strategies for doubling two-digit numbers, using their work with pattern blocks as a context. The class makes a list of all the addition combinations with a sum of 10. Then students make a set of Addition Cards, which they will use as they develop strategies for learning the addition combinations to 10 + 10. Students sort their cards into two sets—the combinations they know, and the combinations they need to learn. The class discusses possible strategies for learning difficult combinations. Students' work focuses on:

■ developing strategies for addition

■ using known addition combinations to help learn unfamiliar ones

Materials

- Completed Student Sheet 6 from Session 1
- Containers of interlocking cubes
- Chart paper
- Addition Cards, Sheets 1–3 (1 each per student)
- Scissors (1 per student)
- Envelopes or resealable plastic bags (1 per student)
- Student Sheet 7 (1 per student, homework)

Activity

Ask students to volunteer how many blocks were in the first half of the designs they recorded on Student Sheet 6. Make a list of these on the board.

What did you do to figure out the total number of blocks from half the blocks?

Students may suggest a variety of methods, such as counting the blocks by 2's, multiplying by 2, or adding the number to itself.

Choose a number from the list, and ask students how they would find the total from that number. You might want to give them a few minutes to talk to each other before they share their strategies.

The purpose of this discussion is to begin to encourage students to use informal strategies based on good number sense, and at the same time discourage them from applying rote procedures.

If students do not have much experience developing and explaining their own strategies, you will have to model good strategies for them. In particular, emphasize looking at the whole problem, estimating, using left-to-right strategies, and using important landmarks in the number system. The **Teacher Note,** Two Powerful Addition Strategies (p. 30), provides more information on addition strategies we want to help students develop.

Continue the discussion, using one or two more examples from the list.

Discussion: Doubling Two-Digit Numbers

What Combinations Make 10?

Distribute the containers of interlocking cubes so that everyone has easy access to them. Working in pairs, students use the cubes to show all the combinations of two numbers that add to 10. Ask them to use one color for one number and another color for the second number, and to put the two groups together to check for 10 cubes. For example, to show 5 + 5, make one group of 5 with red cubes, one group of 5 with blue cubes, and attach them to make 10 (you may want to demonstrate this).

Circulate around the class and notice how fluent your students are with the combinations that make 10. Do they seem to find them all readily? Can they find them systematically?

When most students have made several combinations, list on the board, in whatever order students suggest, all the combinations they have found.

Remember we talked about how mathematicians often put things in order, so they can see if there are any patterns, the way we did with the list of doubles. Let's put this list in order. Which should I put first?

Following students' suggestions, write the list in order on a sheet of chart paper, under the title *Combinations of 10*. (The starting point of the order may vary from class to class—some will start with 10 + 0 or 9 + 1 and continue in descending order, others with 0 + 10 or 1 + 9 and continue in ascending order; either is fine.) Ask students to describe anything they notice about the list and to contribute any missing combinations. Students may raise the issue of whether 2 + 8 is a different combination from 8 + 2. Ask students to explain why they think they are the same or different.

Post this list near the list of doubles for reference throughout the unit. You can tell students that they will be working on learning addition combinations they don't already know, and that they will use the lists of doubles and combinations of 10 to help them with harder combinations.

If your students seem unfamiliar with the combinations of two numbers that make 10, make this a focus of some of your Ten-Minute Math sessions.

Combinations of 10

0 + 10 = 10
1 + 9 = 10
2 + 5 = 10
3 + 7 = 10
4 + 6 = 10
2 + 8 = 10

Give each student a copy of the Addition Cards, Sheets 1–3 (pp. 108–110), an envelope or resealable plastic bag, and a pair of scissors to cut the cards apart (unless you or students have prepared the Addition Cards in advance, in which case simply distribute a set to each student).

Students sort their cards into two piles: "Combinations I know" and "Combinations I am working on." They should mark the back of the Addition Cards they know with an identifying symbol and with their initials. (They should *not* mark the backs of the other cards, as they will gradually mark these as they learn them.)

Some students may prefer to work alone; others in pairs. To work in pairs, one student goes through the cards and asks the other student each addition problem. If the student knows it immediately, it goes in the "Combinations I know" pile; otherwise, it goes in the pile "Combinations I am working on." Then students reverse roles. If there is any doubt about which pile a card goes in, encourage students to call it a "Combination I am working on."

Ask students to volunteer what they think are the most difficult addition combinations, or choose a few you have noticed that many students put in their "Combinations I am working on" pile. You might also choose addition combinations that were (or still are) hard for you—students are always fascinated to learn that adults have difficulty with some of the same things they do. (The hardest combinations for one of the *Investigations* authors have always been 7 + 5 and 8 + 5.)

Put one of these difficult combinations on the board, and ask students to contribute all the strategies they can think of that would help find the total. Do this with several addition combinations.

If your students have been trying to learn by rote rather than with strategies, they may not have much to contribute. Draw their attention to the lists of doubles and combinations that make 10. Ask them how some of these combinations could help them with harder combinations. If students offer no ideas, you may need to model one or two:

Some people think about it this way: Suppose you already know that 5 + 5 equals 10, but you don't know what 5 + 6 is. Can anyone think of a way to use 5 + 5 to help you figure out 5 + 6?

If you know that 6 + 4 adds to 10, could it help you figure out 7 + 4? Who has a way to figure out 7 + 5?

See the **Teacher Note,** Strategies for Learning Addition Combinations up to 10 + 10 (p. 22), for a complete discussion about helping students learn these addition combinations.

Finding Clues to Help with Difficult Combinations   Draw the following Addition Card on the board:

```
┌─────────────────────────────────────┐
│              8 + 6                   │
│              6 + 8                   │
│                                      │
│   Clue: _____    │
└─────────────────────────────────────┘
```

Who has a way to help figure this out?

Take suggestions, then show how to write these statements as clues on the Addition Card. Students need not write their entire thinking as a clue, but just something like a related combination to serve as a reminder. Here are some examples:

I think of 6 + 6 is 12, and then it's 2 more, so it's 14.

I think of 8 and 2 is 10, so take 2 away from the 6 and there's 4 left, so it's 14.

```
┌──────────────────────────┐  ┌──────────────────────────┐
│          8 + 6           │  │          8 + 6           │
│          6 + 8           │  │          6 + 8           │
│  Clue: __6 + 6____       │  │  Clue: __8 + 2 = 10__    │
└──────────────────────────┘  └──────────────────────────┘
```

I know 6 and 4 makes 10, with 4 left, so it's 14.

I just think of 8 + 8, and it's 2 less.

```
┌──────────────────────────┐  ┌──────────────────────────┐
│          8 + 6           │  │          8 + 6           │
│          6 + 8           │  │          6 + 8           │
│  Clue: __6 + 4 = 10__    │  │  Clue: __8+8__           │
└──────────────────────────┘  └──────────────────────────┘
```

Tell students they'll be deciding on clues for some of their own Addition Cards during the next few sessions. Their homework will help them think about good clues to use for addition combinations.

 **Homework**

Strategies for Addition   Give each student a copy of Student Sheet 7, Strategies for Addition. For homework, students write down two of the addition combinations that are hard for them and describe a strategy for learning each one.

....

❖   **Tip for the Linguistically Diverse Classroom**   On Student Sheet 7, tell students to put a question mark next to the sentences that say "I need to work on." Explain that this will remind them where to write combinations that they don't know yet. Have them put a star next to the sentences that say "Here is a strategy that helps" to remind them that this is where they will write clues—things they already know that will help them figure out the problem. Students can show their strategies without writing any words. For example:

?   I need to work on 5 + 4.

⭐   Here is a strategy that helps: 5 + 5 = 10, 5 − 1 = 4, 5 + 4 = 10 − 1 = 9

....

Most of us who are teaching today learned to add starting with the ones, then the tens, then the hundreds, and so on, moving from right to left and "carrying" from one column to another. This algorithm is certainly efficient once it is mastered. However, there are many other ways of adding that are just as efficient, closer to how we naturally think about quantities, connect better with good estimation strategies, and generally result in fewer errors.

When students rely on memorized rules and procedures that they do not understand, they usually do not estimate or double-check. They can easily make mistakes that make no sense. We want students to think about the quantities they are using and what results to expect. We want them to use their knowledge of the number system and important landmarks in that system. We want them to break apart and recombine numbers in ways that make computation more straightforward and, therefore, less prone to error. Writing problems horizontally rather than vertically is one way to help students focus on the whole quantities.

The two addition strategies discussed here are familiar to many competent users of mathematics. Your students may well invent others. Every student must be comfortable with more than one way of adding, so that an answer obtained using one method can be checked by using another. Anyone can make a mistake in routine computation, even with a calculator. What is critical, when accuracy matters, is that you have spent enough time estimating and double-checking to be able to rely on your result.

Left to Right Addition: Biggest Quantities First When students develop their own strategies for addition from an early age, they usually move from left to right, starting with the bigger parts of the quantities. For example, when adding 27 + 27, a student might say "20 and 20 is 40, then 7 and 7 is 14, so 40 plus 10 more is 50 and 4 more makes 54." This strategy is both efficient and accurate. Some people who are

extremely good at computation use this strategy, even with large numbers.

One advantage of this approach is that when students work with the largest quantities first, it's easier to maintain a good sense of what the final sum should be. Another advantage is that students tend to continue seeing the two 27's as whole quantities, rather than breaking them up into their separate digits and losing track of the whole. When using the traditional algorithm (7 + 7 is 14, put down the 4, carry the 1), students too often see the two 7's, the 4, the 1, and the two 2's as individual digits. They lose their sense of the quantities involved, and if they end up with a nonsensical answer, they do not see it because they believe they "did it the right way."

Rounding to Nearby Landmarks Changing a number to a more familiar one that is easier to compute is another strategy that students should develop. Multiples of 10 and 100 are especially useful landmarks for students at this age. For example, in order to add 199 and 149, you might think of the problem as 200 plus 150, find the total of 350, then subtract 2 to compensate for the 2 you added on at the beginning.

To add 27 and 27, as in the previous example, some students might think of the problem as 30 + 30, then subtract 3 and 3 to give them the final result. Of course there are other useful landmarks, too. Another student might think of this problem as 25 + 25 + 2 + 2. There are no rules about which landmarks in the number system are best. It simply depends on whether using landmarks helps you solve the problem.

If your students have memorized the traditional right-to-left algorithm and believe that this is how they are "supposed" to do addition, you will have to work hard to instill some new values— that looking at the whole problem first and estimating the result is critical, that having more than one strategy is a necessary part of doing computation, and that using what you know about the numbers to simplify the problem leads to procedures that make sense.

Finding Doubles and Halves

Materials

- Choice List (1 per student, optional)
- Pattern blocks (several buckets)
- Student Sheet 5 (10 copies to share)
- Student Sheet 6 (1 per student)
- 100 charts (4–5 per student)
- Numeral Cards (1 deck per pair)
- Crayons or markers
- Doubles and Halves Problems (5–6 sets, or booklet for each student)
- Student Sheet 8 (2–3 per student, 1 for homework)
- Students' Addition Cards
- Interlocking cubes (available as needed)
- Addition Cards, Sheets 1–3 (1 set per student, homework)

What Happens

Students are introduced to Choice Time, during which they choose from among several activities: working with symmetrical pattern block designs, playing Plus–Minus–Stay the Same, working out problems about doubles and halves, and learning some of the addition combinations they found difficult earlier in the investigation. Their work focuses on:

- developing strategies for addition problems
- exploring which numbers can be divided in half evenly
- learning new addition combinations by reasoning from known combinations

 Ten-Minute Math: Calendar Math   Continue using Calendar Math as a short activity outside of math class. You might want to introduce the following constraint: Use a double or a combination that makes 10 as part of your number expression. For example, if today is September 15, possible equivalents for 15 using doubles or tens combinations include: $7 + 7 + 1$, $8 + 8 - 1$, $7 + 3 + 5$, $5 + 5 + 5$, and $10 + 10 - 5$.

Activity

Choice Time: Working with Doubles

Introducing Choice Time   Tell students that the next few math sessions will be Choice Time, and they will have the choice of working on several different activities.

Choice Time is a format that recurs throughout the *Investigations* units, so this is a good time for students to become familiar with its structure. They had a little experience with this format in Investigation 1 when they had a choice of working on their 100-cube construction or playing the Plus–Minus–Stay the Same game. Now, with four choices, they will need more structure in which to make and keep track of their choices. See the **Teacher Note,** About Choice Time (p. 36), for information about how to set up Choice Time, including how students might use the Choice List (p. 113) to keep track of their work.

Introduce your rules and students' responsibilities for Choice Time. During each one-hour class session, students can participate in one or two of the choices. Students can select the same activity more than once, but they should not do the same one every day. They don't have to do every activity.

Many teachers have their students keep a record of what they do each day—on a piece of paper, in their mathematics notebooks, or on the Choice List (p. 113). Or, you may want to set up a large classroom chart, with student names down one side and the choices across the top, on which students mark the activities as they do them. On such a chart, the boxes should be large enough to accommodate several checks, since students may do any one activity more than once.

	Pattern Block Designs	Plus-Minus Stay the Same	Doubles and Halves	Learning Addition
Jennifer	✓			✓
Yoshi		✓✓	✓	
Dylan	✓	✓		
Kate	✓	✓	✓	✓
Maya		✓	✓✓	
Dominic	✓✓			✓
Latisha			✓	✓✓
Aaron	✓		✓	
Jamal		✓		✓

How to Set Up the Choices   If you set up your choices at centers, show students what they will find at each center. Otherwise, make sure they know where they can find the materials they need.

Choice 1: Pattern Block Designs—copies of Student Sheet 5 (to be shared), copies of Student Sheet 6 for each student, pattern blocks

Choice 2: Plus–Minus–Stay the Same—copies of 100 charts, decks of Numeral Cards, crayons or markers for covering numbers

Choice 3: Doubles and Halves Problems—sets of problems assembled as booklets or collected in envelopes, copies of Student Sheet 8 for each student

Choice 4: Learning Addition Combinations—students' own sets of Addition Cards

Cubes and 100 charts should be available for use with any of these activities. For example, you may want to encourage (or even insist) that students get cubes and charts for figuring out doubling problems. Remind students that they can also use the classroom chart of doubles to help them in these activities.

Introduce the Four Choices Remind students how to do Pattern Block Designs (introduced in Session 1, p. 19), and how to play Plus–Minus–Stay the Same (introduced in Investigation 1, p. 11). Briefly explain the other two choices, Doubles and Halves Problems and Learning Addition Combinations; see the following activity descriptions for specifics.

Choice 1: Pattern Block Designs

As they did earlier, students build half of a pattern-block design on Student Sheet 5, Half and Half; count the number of pattern blocks they used; predict how many they will need to make the whole design; then complete the design and count the total number of blocks. Students may work in pairs or individually, but each student should record on his or her own copy of Student Sheet 6, How Many Blocks? As you watch students work, continue to help them write and draw about their strategies for predicting. Emphasize double-checking their work.

❖ **Tip for the Linguistically Diverse Classroom** For students with limited English proficiency, follow the suggestions on p. 19 for working with Student Sheet 6.

Choice 2: Plus–Minus–Stay the Same

Students play this game with a partner (refer to the rules on p. 99 as necessary). You can introduce the following variation to small groups if you feel they are ready to deal with increased complexity: In addition to adding 10, subtracting 10, or using the same number they turned over, players may add 20, subtract 20, or double the number. Provide a list of these new options for anyone who is ready. Many students will need more experience playing the basic game without additional options.

Choice 3: Doubles and Halves Problems

Students choose problems from the sets of Doubles and Halves Problems. Some teachers put the problems into a booklet for each student; students don't necessarily do every problem, but choose one or two each time they work on it. The booklets can also be sent home (or, you may want to pull out one or two of the problems to use specifically as a homework assignment). Other teachers make several sets of the cut-apart problems available in boxes or envelopes. Students take a set to their seat, then select a problem to work on.

Students record their work on Student Sheet 8, Problem Strategies, writing which problem they solved (for example, Doubles and Halves Problem 1), stating the problem in their own words, and recording how they solved it, using words, pictures, and numbers.

❖ **Tip for the Linguistically Diverse Classroom** Read aloud (or ask an English-proficient student to read aloud) each problem before students try to solve it. Allow students to complete Student Sheet 8 in their native language. If they have not acquired such writing skills, they might show you (or a partner) how they solved the problem.

As you observe students working, you may want to adjust the difficulty of the numbers in the problems for different students. Often, students will self-regulate, choosing problems of appropriate difficulty. You can encourage this by saying something like, "Choose a problem that's not too easy or too hard. Find one that feels medium."

If you notice that some students tend to choose problems that are too difficult or too easy, you might choose another problem for them or modify the numbers in the problem: "I'd like you to try this with some different numbers. Let's say that there are 11 birds (instead of 29)." See the **Teacher Note,** Collaborating with the Authors (p. 37), for more about your role and your students' roles in finding the right level of challenge for each of them.

Choice 4: Learning Addition Combinations

Students select between five and ten of their Addition Cards from the "Combinations I am working on" sets. They add clues to these cards, thinking of combinations they already know that will help them learn these. Students may work in pairs to brainstorm clues; then students write on their card the clue that works best for them.

You may want to meet with groups of students to help them get started selecting clues. Let them know that different clues will work best for different people; the clues that help one student might be different from the clues a partner chooses.

After writing clues on all the cards they have chosen, pairs work together on learning the combinations. They trade cards and take turns reading them to one another, giving one of the combinations on the card and pausing while the partner tries to give the total. The pairs read through all their cards several times, sometimes reading one form of the fact (say, 5 + 7), sometimes the other (7 + 5). Whenever a student needs a hint, the partner reads the clue written on the card.

Monitoring Student Work   For the rest of this investigation (Sessions 3–7), students will be working independently on Choice Time activities, with a few breaks for whole-group discussions and assessment activities. There are more problems and activities than any student will complete in this time. You will want to make sure that all students spend some time on the Money Problems, after they are introduced in Session 5, since these problems provide important diagnostic information.

Additionally, all students should spend some time writing clues and practicing with their Addition Cards. Even those students who already know all of the addition combinations should spend some time writing clues. For them, this activity will provide an opportunity to focus on categories, such as doubles and tens, and to explore how combinations are related to each other.

Sessions 3 and 4 Follow-Up

Problem Strategies   Students select one or two of the Doubles and Halves Problems they have not done during Choice Time to do at home. They can copy the problem onto Student Sheet 8, Problem Strategies, or you might staple the problem to the student sheet.

 **Homework**

You may want to have students work on their Addition Cards at home. However, since they need to have a set of the cards in school and they may not remember to bring them back, they should make a home set from additional copies of Addition Cards, pages 1–3 (pp. 108–110).

If possible, give students time to cut out their home set at school, and provide an envelope or plastic bag for storage. At home, students can continue to select clues for the combinations they are trying to learn. A family member could then give them the problems for practice. To check their answers, students might use something they have at home to count with, such as dried beans, pennies, checkers, or paper clips.

Difficult Addition Combinations   You may want to make a classroom list of addition combinations that are difficult for many students. Next to each one, you could list some of the clues students have chosen to help them with these combinations. Students enjoy knowing that some of their problem combinations are also difficult for many others. Again, you might share any that are still difficult for you as an adult.

 **Extension**

Choice Time is an opportunity for students to work on a variety of activities that focus on similar content. The activities are not sequential; as students move among them, they continually revisit some of the important concepts and ideas they are learning—for example, in Sessions 3–7, concepts related to adding numbers, doubling, and halving. Some activities require that students work in pairs, while others can be done either alone or with a partner. Most activities require some type of recording or writing; these records will help you assess students' growth.

Students can use the Choice List (p. 113) to keep track of their work. As students finish a choice, they write it on their list and attach any work they have done. Some teachers list the choices for each day on the board or overhead and have students copy this list at the beginning of class. Students are then responsible for checking off completed activities. You may also want to make the choices available at other times during the day.

In any classroom there will be a range of how much work students can complete. Each choice may also provide extensions and additional problems for students to do once they have completed their required work. Choice Time encourages students to return to choices they have done before, doing another page of problems or playing a game again. Students benefit from such repeated experiences.

If you and your students have not used a structure like Choice Time before, establish clear guidelines when you introduce it. Discuss what students' responsibilities are during Choice Time:

- Try every choice at some time.
- Be productively engaged during Choice Time.
- Work with a partner or alone.
- Keep track, on paper, of the choices you have worked on.
- Keep all your work in your math folder.
- Ask questions of other students when you don't understand or feel stuck.

Some teachers establish the rule, "Ask two other students before me," requiring students to check with two peers before coming to the teacher for help.

You may need to try organizing Choice Time in a couple of different ways and decide from experience which approach best matches the needs of your students.

CHOICE LIST

Name *Saloni*

Activity Choice   ✔ When Finished

1. pattern block design ✔
2. plus minus stay the same ✔
3. learning addition ✔
4. money problems ✔
5. Doubles and Halves ✔
6. plus minus stay the same ✔
7. plus minus stay the same ✔
8. _____ ☐

Collaborating with the Authors

Every unit in this curriculum is a guide, not a prescription or recipe. We tested these activities in many different classrooms, representing a range of students and teachers, and revised our ideas constantly as we learned from students and teachers alike. Each time we tried a curriculum unit in a classroom, no matter how many times it had been tried and revised before, we discovered new ideas we wanted to add and changes we wanted to make. This process could be endless, but at some point we had to decide that the curriculum worked well enough with a wide range of students.

We cannot anticipate the needs and strengths of your particular students this particular year. We believe that the only way for a good curriculum to be used well is for teachers to participate in continually modifying it. Your role is to observe and listen carefully to your students, to try to understand how they are thinking, and to make decisions, based on your observations, about what they need next. Modifications to the curriculum that you will need to consider throughout the year include the following:

- changing the numbers in a problem to make the problem more accessible or more challenging for particular students

- repeating activities with which students need more experience

- engaging students in extensions and further questions

- rearranging pairs or small groups so that students learn from a variety of their peers

Your students can help you set the right pace and level of challenge. We have found that, when given choices of activities and problems, students often do choose the right level of difficulty for themselves. You can encourage students to do this by urging them to find problems that are "not too easy, not too hard, but just right." Help students understand that doing mathematics does not mean knowing the answer right away. Tell students often, "A good problem for you is a problem that makes you think hard and work hard—and you might have to try more than one way of doing it before you figure it out."

The *Investigations* curriculum provides more than enough material for any student. Suggestions are included for extending activities, and some curriculum units contain optional sessions (called Excursions) to provide more opportunities to explore the big mathematical ideas of that unit. Many teachers also have favorite activities that they integrate into this curriculum. We encourage you to be an active partner with us in creating the way this curriculum can work best for your students.

Doubling with Money

Materials

- All the Choice Time materials from Sessions 3–4
- Money Problems (5–6 sets)
- Plastic coins (small collection per pair)
- Cubes and 100 charts
- Overhead projector
- Student Sheet 8 (2 per student)
- Student Sheet 9 (1 per student, homework)
- Student Sheet 10 (1 per student, homework)
- Student Sheet 11 (1 per student)
- Student Sheet 12 (1 per student, homework)
- Student Sheet 5 (1 per student, homework)

What Happens

Choice Time continues after students participate in an introductory activity with coins, which is also a Checkpoint for you to look at their familiarity with coin values. A new choice, Money Problems, is added to Choice Time. Students discuss and compare strategies for doubling problems. In an assessment activity, students solve a two-digit addition problem in two ways and record their work. Their work focuses on:

- recognizing the values of coins
- finding the value of a collection of coins
- combining collections of coins
- developing addition strategies
- learning addition combinations

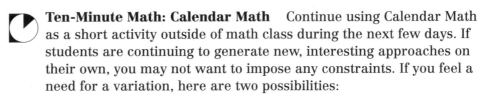 **Ten-Minute Math: Calendar Math**   Continue using Calendar Math as a short activity outside of math class during the next few days. If students are continuing to generate new, interesting approaches on their own, you may not want to impose any constraints. If you feel a need for a variation, here are two possibilities:

Coin Values   Since students will be dealing with the coin values of 5, 10, and 25, have them include as many of those values as they can in their equivalent expressions for the date (or, to make it harder, have them use only coin values). For example, if today is September 16, possible equivalents are $5 + 10 + 1$, $25 - 10 + 1$, and $(5 \times 5) - 10 + 1$.

Calculator Equivalents   Each student or pair needs a calculator. You give them a number they must start with; for example, 100, 75, or 50. They find equivalent expressions for the date starting with this number and using a calculator to experiment. Students should have paper and pencil to record what they do on the calculator. For example, if today is September 16 and the starting number is 100, possible equivalents are $100 - 50 - 30 - 4$, $100 - 100 + 16$, and $100 - 90 + 6$. This constraint pushes students to use subtraction and to use the calculator as a reasoning tool.

Counting Money

Hand out a small collection of coins to each pair. Ask students what they know about the names and values of the various coins. Then do a quick check of how familiar students are with them.

Hold up the nickel. Hold up the coin that's worth 10 cents. Hold up the coin that's worth the most. Hold up the coin that's worth 1 cent. Hold up coins that make 15 cents altogether.

Ask students to figure out how much money they have in their collection. Make sure they know they can use cubes or the 100 chart to help them. Since students may not be used to having these materials available, you may need to insist that they use cubes or 100 charts to help them find the total or to double-check their work.

When each pair has a total, they write it on a slip of paper so they don't forget it, then trade coin sets with another pair. Each pair now finds the total of the new collection and compares their answer with the first pair's answer. If they don't agree, each pair demonstrates to the other how they solved the problem, and they try to reach a consensus.

As students are working, observe their familiarity with coins and with combinations of coins:

- Do they know the values of the coins?
- Do they know how to count by 5's, 10's, 25's?
- Are they successful at combining coins of like denominations (for example, all the nickels, all the dimes)? Of unlike denominations (for example, a dime and a quarter)?
- Are they able to use materials—cubes or 100 charts—appropriately to help them solve problems?
- Are they using mental strategies based on their knowledge of the numbers?

If you notice that some sets of coins are too easy or too difficult for some students, modify their collections on the spot by adding or taking away some of the coins.

Choice Time: Doubles and Halves

During the rest of Session 5 and for parts of Sessions 6 and 7, students continue working on Choice Time. Remind them to keep track of which activities they work on. They choose from among the four choices described in Sessions 3–4, and one new choice:

Choice 1: Pattern Block Designs

Choice 2: Plus–Minus–Stay the Same game (played in pairs, with or without the added options)

Choice 3: Doubles and Halves Problems

Choice 4: Learning Addition Combinations

Choice 5: Money Problems—provide plastic coins, sets of Money Problems, interlocking cubes and 100 charts, and copies of Student Sheet 8, Problem Strategies, for each student.

Choice 5: Money Problems

Introduce the new choice, Money Problems. Students are to take a set of Money Problems to their seat and select one. On Student Sheet 8, Problem Strategies, they write which problem they solved (for example, Money Problem 1), state the problem in their own words, and record how they solved it, using words, pictures, and numbers.

❖ **Tip for the Linguistically Diverse Classroom** For students who have difficulty reading English, read aloud (or ask another student to read aloud) the selected Money Problem. Students then write the number of the problem and illustrate what it asks on Student Sheet 8. They may use mostly pictures and numbers to show how they solved the problem.

Tell students you would like everyone to work on a few Money Problems sometime during the remainder of this session and the next two class sessions, but to do no more than one or two problems each day. Encourage them to choose problems that are not too difficult and not too easy.

As you watch, encourage students to use interlocking cubes or the 100 chart to model addition; the coins themselves are helpful in some ways, but do not help students think about the structure of their values. For example, suppose a student has a dime and a quarter and is trying to double that amount. The student might take two stacks of 10 cubes for each of the dimes, and two and a half stacks for each of the quarters. The cubes help the students see how to count by 10's, and how two 5's make another ten.

While Students Are Working on the Choices

■ Leave time at the end of Session 6 for the next activity, a whole-class discussion, Doing Doubles Problems. Also refer to p. 44 for suggested homework after Session 6.

■ Before Session 7, you may want to review the records to see which activities students have worked on. Make sure that all students have tried at least two choices, including the Money Problems.

■ Plan to spend the last half of Session 7 on the assessment task, Writing About Addition (p. 43).

Discussion: Doing Doubles Problems

Devote about 20 minutes to this discussion at the end of Session 6. Have interlocking cubes within reach of all students. Pose one doubling problem similar to those students have been doing, but change the numbers to create a problem of medium difficulty.

❖ **Tip for the Linguistically Diverse Classroom** Pose problems using vocabulary that refers to objects that can be pointed to or easily drawn on the board.

Here are a few suggestions:

I saw 15 birds sitting on a wire. How many wings did all those birds have?

I have 24 plants in my garden. My neighbor has twice as many plants. How many plants does she have?

One third grade class has 26 students. The other third grade class has the same number. How many third graders are there?

Write the number in the problem on the board or overhead for students to refer to as they work. Pairs work on this problem for a few minutes. When many seem to have a solution, ask students to share their strategies. Record these in some way to model for students how they might record their solutions. Point out ways in which students have used interlocking cubes or drawings as they worked.

Do another doubling problem in this same way.

Throughout the discussion, emphasize and support good strategies based on number sense, including left-to-right addition and rounding to familiar landmarks. For more about the kinds of numerical strategies we want to help students develop, see the **Teacher Note,** Two Powerful Addition Strategies (p. 30).

If a student says, "It's 26 and 26, so 6 and 6 is 12, put down the 2 and carry the 1," ask, "Can you think about it by using doubles?" or, "Can you show me how that works with cubes?" For more concerns about students who use only a memorized procedure, see the **Teacher Note,** Developing Strategies for Addition (p. 45).

Writing About Addition

At the end of Session 7, give a copy of Student Sheet 11, Addition in Two Ways, to each student.

Most of the time I want you to work together, but today I'd like to see how each of you thinks about this problem by yourself. I'm going to look at all the different ways people in this class can come up with for solving this problem.

Please try to find two ways to solve it so you can double-check yourself. Use cubes or 100 charts, make drawings, or do anything else that will help you solve the problem.

Observe what students are doing as they solve the problem. If you keep notes about students' work, this is a good time to jot down observations.

- Do students count by 1's, or do they group by 2's, 5's, or by 10's and 1's?
- Do students have good strategies for organizing and keeping track of their counting and adding?
- Do students apply the addition combinations they know?
- How do students add 20 and 20? Do they build it first? Do they count by 10's? Do they know that 20 and 20 equals 40 without counting?

You may have a few students for whom you have to modify this problem. You can use Student Sheet 12, More Addition in Two Ways, to write different numbers for this problem. If a student is struggling to count by 1's and can't keep track of the numbers, or has no sense of what the quantity 27 is, give the student a problem with two smaller numbers, such as 13 and 15. The fact that you need to modify the problem is important diagnostic information.

When you have collected the papers, look at them with the following questions in mind:

- Can students think of two ways to solve the problem?
- Can students record their work clearly, using numbers, words, and pictures to communicate what they did?

If you are keeping student portfolios, you may want to include this work.

As you finish Investigation 2, be sure students keep their Addition Cards for continued practice. They will also need them again in Investigation 4, when they will use them for reference as they explore the characteristics of odd and even numbers.

Sessions 5, 6, and 7 Follow-Up

 **Homework**

Money Problems to Do at Home   After Session 5, send home Student Sheet 9, Money Problems to Do at Home. Plan to review the completed sheets to check on students' familiarity with coins and their values.

❖ **Tip for the Linguistically Diverse Classroom**   To prepare for work on Student Sheet 9, read each problem aloud, modeling actions as you do so. For example, when reading aloud problem 3, show a purse and remove all coins; show your pockets and count the change.

How Much Is Your Symmetrical Design Worth?   After Session 6, send home Student Sheet 10, How Much Is Your Symmetrical Design Worth? and another copy of Student Sheet 5, Half and Half. Plan to review the completed sheets to check on students' familiarity with coins and their values and students' addition strategies.

More Addition in Two Ways   After Session 7, send home Student Sheet 12, More Addition in Two Ways, in which students choose their own numbers. Look over the completed homework for more information about students' addition strategies and to see whether students choose numbers that provide the right level of challenge for themselves.

Developing Strategies for Addition

By third grade, students should be developing several different procedures that they can use fluently and flexibly to solve addition problems. Some students will come to your class with previous experience in developing their own procedures based on what they know about operations and number relationships. Others may know only the historically taught column addition, which uses carrying.

Help students develop their own strategies based on what they know about the numbers and operations involved in the problem. Encourage students to make an estimate first, to think about the 10's and 1's in the problem and to notice what landmark numbers (such as multiples of 10 or 25) the numbers in the problem are close to. Suppose the problem involves adding up three 17's (17 + 17 + 17). Many competent math users—children as well as adults—solve this problem by adding the tens first:

> 10, 20, 30, then two 7's is 14, so 30 and 10 and 4 is 44, then 7 more is . . . let's see, 6 more would be 50, so it's 51.

Or they might use nearby landmarks:

> Well, 17 is close to 20, three 20's is 60, then subtract 3 for each 17, so minus 9 is 51.

Of course, even when using commonsense methods, students may still make errors and must still check for accuracy, but if they use methods that make sense to them, they are more likely to be able to find their own mistakes.

It's fine for students to use the traditional addition algorithm, which relies on "carrying," if they understand clearly how and why they are using it. Students who simply use this procedure without thinking about the numbers in the problem may make the following error:

$$
\begin{array}{r}
{}^{1}17 \\
17 \\
+17 \\
\hline
42
\end{array}
$$

These students are saying, "Put down the 2 and carry the 1." In their focus on the mechanics of this rule, they fail to see that they should instead "Put down the 1 and carry the 2." To see if students understand how their addition procedures model the situation, ask questions like these:

Can you show me how that would work with the cubes or on the 100 chart?

How would you show a first grader who doesn't understand what you just wrote how you solved this problem?

Did you make an estimate? Is your answer close to your estimate? What other method can you use to double-check your solution?

Throughout this unit, we stress having students build up their knowledge of number relationships so that they can use the number relationships they know to solve complex problems. For more information on approaches that make sense to students, see the **Teacher Note,** Two Powerful Addition Strategies (p. 30).

Data and Handfuls

What Happens

Sessions 1 and 2: Collecting and Representing Data Students observe, classify, count, and record data about themselves. They play a game called Guess My Rule as a way of collecting information, and then work in pairs to display that information using graphs, pictures, or models. They help construct a bar graph to represent the categories within a set of data.

Sessions 3 and 4: Handfuls of Cubes and Other Objects Students determine how many objects—such as beans, cubes, or tiles—they can hold in each of their hands. They count the objects in two different ways. They combine the amounts they can hold in their hands and then compare the amounts they can hold in their right and left hands. They discuss and record their strategies for combining and comparing amounts using numbers, words, and sketches.

Mathematical Emphasis

- Sorting and classifying information
- Collecting, recording, and representing data
- Describing data presented in tallies or graphs
- Using grouping to count
- Developing strategies to combine and compare quantities

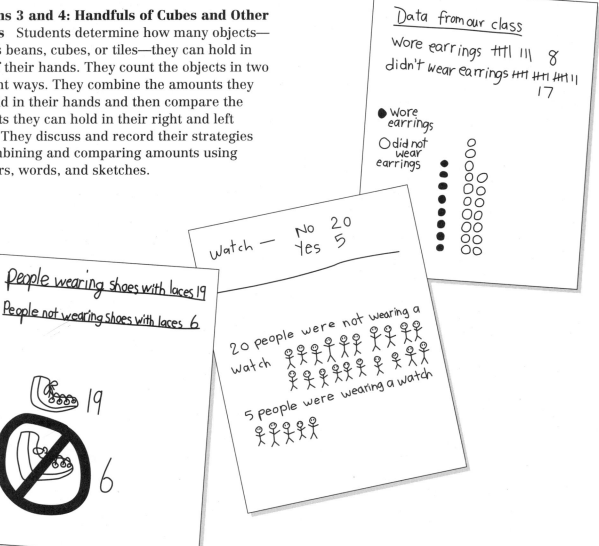

What to Plan Ahead of Time

Materials

- Interlocking cubes (all sessions)
- Paper: plain, lined, and graph (Sessions 1–2)
- Chart paper (Sessions 1–2)
- Colored pencils, markers, or crayons (Sessions 1–2)
- Overhead projector
- Calculators: 1 per student or pair (Sessions 3–4)
- Open containers of uniform items large enough to grab by handfuls (including cubes, but also kidney or lima beans, square tiles, pennies, buttons, or similar items): at least 4 (Sessions 3–4)

  Note: Choose objects of various sizes to allow flexibility for students who need to work with smaller numbers and those who can handle counting, combining, and comparing larger numbers. The bigger the objects, the fewer the number in the handful. Shoe boxes or plastic tubs work well as containers. The more containers of items available, the easier it will be for students to take turns grabbing handfuls to count.

Other Preparation

- Read the **Teacher Note,** Playing Guess My Rule (p. 56), to get ready for the game.

  Duplicate student sheets and teaching resources (located at the end of this unit) in the following quantities. If you have Student Activity Booklets, no copying is needed.

For Sessions 1–2

Student Sheet 13, Guess My Rule (p. 116): 1 per student

One-centimeter graph paper (p. 121): 1–2 per pair

Student Sheet 14, Data Analysis (p. 117): 1 per student (homework)

Student Sheet 15, Calendar Math (p. 118): 1 per student (homework)

For Sessions 3–4

Student Sheet 16, Handfuls (p. 119): 3–4 per student

Student Sheet 17, Handfuls at Home (p. 120): 1 per student (homework)

Collecting and Representing Data

Materials

- Interlocking cubes
- Paper: plain, lined, and graph
- Chart paper
- Colored pencils, markers, or crayons
- Student Sheet 13 (1 per pair)
- Student Sheet 14 (1 per student, homework)
- Student Sheet 15 (1 per student, homework)
- Overhead projector

What Happens

Students observe, classify, count, and record data about themselves. They play a game called Guess My Rule as a way of collecting information and then work in pairs to display that information using graphs, pictures, or models. They help construct a bar graph to represent the categories within a set of data. Their work focuses on:

- collecting information about a group of people
- sorting and classifying information
- counting and comparing sets of data
- using pictures, tallies, and graphs to organize and display data

Activity

Playing Guess My Rule

Introduce this investigation by telling students that, as part of their work in mathematics this year, they will sometimes collect information about themselves and their families or about groups of people in the school. They will figure out ways to organize and describe the information they collect. One of the ways they will organize information is by thinking about the way things can be grouped.

When scientists and mathematicians study the world, they often try to think about how things are the same and different. Sometimes things go together one way, but if you think about them differently, you'll see they can go together in a new way.

For example, some people think all third graders go together because they are in the third grade, and some people think certain third graders go with certain second graders or fourth graders because they all play baseball, all read the same kind of books, or all walk to school.

Focus on characteristics of students in your classroom to give other examples about how students might go together in different ways.

Today we are going to play a game called Guess My Rule. You will have to pay attention to a lot of characteristics to figure out how certain groups of people go together. Let's try the game. I will think of a secret rule. Some people will fit my rule, and some people won't. You're going to guess what my rule is.

For these class demonstrations, choose straightforward, visually obvious rules such as HAS BROWN HAIR, WEARING STRIPES, or WEARING SHOES WITH LACES. Tell students you will be grouping them by a characteristic they can observe, such as HAS RED HAIR or WEARING PANTS, and not by a characteristic no one can see, such as LIKES CHOCOLATE ICE CREAM or HAS A DOG. For more information, see the **Teacher Note,** Playing Guess My Rule (p. 56).

Choose two students who fit your rule and have them stand in a designated area where students can see them.

I have a secret rule in mind that tells something about people in this class. It's something you can see. Some people fit my rule and some people do not. Rashad and Ryan both fit my rule. Rashad and Ryan, please go stand by the chalkboard where everyone can see you.

Who thinks they know someone who should stand with Rashad and Ryan? Don't guess what my rule is yet! Right now, just tell me who else you think goes in the group with Rashad and Ryan.

Students take turns saying who they think fits the rule. If the person fits, ask him or her to stand with those who fit the rule. If the person doesn't fit, send him or her to stand in another area designated for those who don't fit.

Stress the importance of all clues—people who fit the rule as well as people who don't. See the **Dialogue Box,** Playing Guess My Rule (p. 58), for an example of how clues are gathered as the game progresses.

When enough good evidence has been gathered and you sense that most students have a good idea about the rule, allow them to guess what it is.

❖ **Tip for the Linguistically Diverse Classroom** Ask students to point to the item they believe dictates the rule. For example, a student points to the shoelaces of everyone in the group standing by the board.

Ask them about the reasons for their conjectures. It is possible that students will suggest categories that do fit the evidence but aren't the ones you had in mind. If this happens, acknowledge the students' good thinking.

Collecting and Recording Data Before students return to their seats, record the data about the number of people who did and did not fit the rule on chart paper, so that it can be easily saved. If some students are still in their seats, have them place themselves into the appropriate group. Have students count the number of people in each group. For example:

WEARING A WATCH: 11

NOT WEARING A WATCH: 15

❖ **Tip for the Linguistically Diverse Classroom** Use simple pictures or other visual aids when recording the data.

Continue playing more rounds of Guess My Rule. Play at least one game where the rule is WEARING SHOES WITH LACES. You will need this information in the next activity. Record the data from each game in a different way, using pictures, tallies, numbers, checkmarks, and so forth to model a variety of ways of keeping track of data. Students may have other suggestions for recording data. Encourage variety and innovation.

Note: The rules we suggest may not always work to classify students in your classroom. For instance, in some schools all students might have the same hair color, or students may all wear school uniforms with buttons. Choose rules that are descriptive of your class.

Spend a few minutes looking at the collected data. Have students figure out the number of students in each group. You might ask them to combine the totals for each rule or to compare the numbers for each group.

How many more people are wearing shoes with laces than aren't wearing shoes with laces?

If there are 26 students in our class and 10 have brown hair, how many people don't have brown hair?

If no one notices, ask students why they think that the total number for each data set is 26 (or whatever the total number of students in your class is).

Discussion: Can Data Change?

Looking at the collected data, ask students to think about which of these sets of data might change from day to day, and which information would not change.

When we played Guess My Rule, we were thinking about ways students in our class can be grouped. Think about the types of information we collected. If we collected this same information tomorrow, would any of the numbers be different? Would any be the same?

Encourage students to express their ideas. Follow up with questions that probe their thinking.

Does anyone agree or disagree with Maria that the number of students in the category HAS BROWN HAIR wouldn't change? Why or why not? Is there any information you are sure would be different? Why do you think that?

Choose a category about which there has been some discussion. Discuss a data set that could change (WEARING SHOES WITH LACES) and one that would probably not change (HAS BROWN HAIR). If students don't suggest it, ask them to consider how the data would change if someone were absent, and which data sets would remain the same as long as all students were in class.

Tell students that tomorrow you will collect the data again to see if the numbers have changed.

Representing the Data from Guess My Rule

Working in pairs, students choose one of the data sets collected during Guess My Rule and find a way to represent the information using interlocking cubes, pictures, numbers, or graphs.

Often when mathematicians collect information, they find ways to show it in different ways, both so they can see it in new ways and so they can share it with other people. Sometimes they make pictures or graphs, and sometimes they build models. They call these *representations* of the data.

With your partner, choose a set of data that we collected during Guess My Rule. Make your own representation of this data. You can use cubes, draw a picture, or make a graph of the data. For example, if you pick WEARING A WATCH, you could draw something so someone who wasn't here in class with us could figure out how many of us wore watches and how many did not.

Before students begin working, have them generate ideas about how they might show the data. The word *representation* is often unfamiliar to third grade students, so use it along with words like *picture, graph, chart,* and *model.* Give an example, like WEARING A WATCH, and solicit ideas about how students might organize a picture or graph about that data, and what materials they might use. See the **Teacher Note,** Sketching Data (p. 57), for more details about what is important for students to do as they make their representations.

Distribute a copy of Student Sheet 13, Guess My Rule, to each pair. Make available a variety of graphing materials (graph paper, interlocking cubes, colored pencils or crayons, and so forth). If students are having difficulty getting started, help them think about the two pieces of information they need to represent and which materials they might want to use.

Observing the Students As students work in pairs, circulate around the room and observe the following:

- How are students organizing the information?
- Are they accurately representing the information?
- What materials and methods have they chosen to represent the data?

Ask students to explain their representations to you.

- Can they interpret their representations?
- Can they extract the important information from their representations?

Asking students to explain their work allows you to learn more about how they are thinking, and often helps them clarify their thinking. As students talk about their work, they will self-correct or clarify aspects that do not make sense. Begin to encourage partners to ask each other questions about their work.

Sharing Representations Have students share their work by asking several groups that worked with the same set of data to bring their representations to the front of the room.

If you chose to represent the information about [wearing stripes], bring your graphs or models up front so everyone can see them.

Comment on the variety of representations and materials used. If you have time, ask each pair to say one thing about their representation. Find a place to display the representations so students can get a closer look at a later time.

Representing Data with Categories

When students organized their Guess My Rule data, they were dealing with two groups: the group of people who fit the rule and the group of people who did not. In some cases, the people in the latter category could not be described any other way. For example, either students ARE or ARE NOT WEARING WATCHES, and the students in the NOT WEARING WATCHES group cannot be further described using the category of watches. In contrast, the students in the group NOT WEARING SHOES WITH LACES *can* further be described by other categories of shoe type. Introduce this idea to your students:

Yesterday we collected some data about the number of people in our class who were wearing shoes with laces and the number of people who were not wearing shoes with laces. This gives us information about the type of shoes worn by some students, but there is a whole group of students for whom the only thing we know about them is that they were NOT wearing shoes with laces. What are some other categories we could use to further describe the types of shoes worn by the people in this second group?

Solicit ideas from students, and list them on the board or overhead. Help them organize their list of categories into a simple bar graph. Ask students who fit each category to stand up. Enter one square, X, or stick figure on the graph for each student.

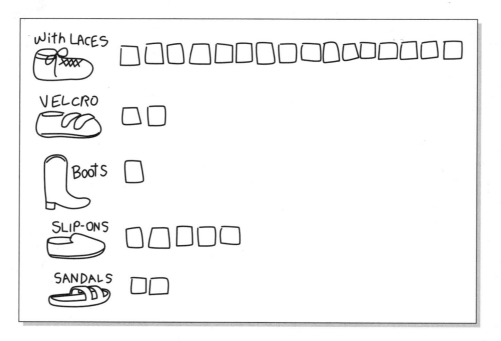

When all the data for your class have been recorded, ask students what types of things they can now tell about shoes worn by the people in their class.

What information does this graph tell us?

What can you say about the number of people wearing shoes with laces compared to the number of people wearing sandals?

How does this representation differ from the representation some of you made yesterday using the Guess My Rule data?

Would this data change if we collected the same information tomorrow? Why or why not?

❖ **Tip for the Linguistically Diverse Classroom** Restructure some of the discussion questions so that they can be answered in one-word responses. For example: How many people are wearing sandals? How many people are wearing shoes with laces? Are more or fewer people wearing sandals? Did the Guess My Rule data tell us what other shoes people were wearing?

Ask students how they might use interlocking cubes to build a model that would represent this shoe data. Have a pair represent the data using cubes.

Students will continue to collect, record, and interpret data in Sessions 3 and 4 of this unit and in other units in the *Investigations* curriculum. Consider integrating data collection and representation into other aspects of your curriculum, such as science or social studies. Collecting data about themselves is a wonderful way for students to get to know each other at the beginning of the school year.

Sessions 1 and 2 Follow-Up

Data Analysis   For practice with representing and analyzing data, children do Student Sheet 14, Data Analysis, after Session 1.

🏠 **Homework**

Calendar Math   After Session 2, students do Student Sheet 15, Calendar Math. They find three ways to make the number that is today's date.

Guess My Rule is a classification game in which players try to figure out the common characteristic, or attribute, of a set of objects. To play the game, the rule maker (who may be you, a student, or a small group) decides on a secret rule for classifying a particular group of things. For example, rules for people might be WEARING BLUE or HAS BROWN HAIR.

The rule maker starts the game by giving some examples of people who fit the rule—for example, by having two students who are wearing blue stand up. The guessers then try to find other individuals who might fit the rule: "Does Saloni fit your rule?"

With each guess, the individual named is added to one group or the other—*does fit* or *does not fit* the rule. Both groups must be clearly visible to the guessers so they can make use of all the evidence—what does and does not fit—as they try to figure out what the rule is.

You'll need to stress two guidelines during play:

"Wrong" guesses are clues and just as important as "right" guesses. "No, Cesar doesn't fit, but that's important evidence. Think about how Mark is different from Kate, Ly Dinh, and Jeremy." This is a wonderful way to help students learn that errors can be important sources of information.

When you think you know what the rule is, test your theory by giving another example, not by revealing the rule. "Midori, you look like you're sure you know what the rule is. We don't want to give it away yet, so let's test out your theory. Tell me someone who you think fits the rule." Requiring students to add new evidence, rather than making a guess, serves two purposes. It allows students to test their theories without revealing their guess to other students, and it provides more information and more time to think for students who do not yet have a theory.

When students begin choosing rules, they sometimes think of rules that are either too vague

(WEARING DIFFERENT COLORS) or too hard to guess (HAS A PIECE OF THREAD HANGING FROM HIS SHIRT). Guide and support students in choosing rules that are "medium hard"—not so obvious that everyone will see them immediately, but not so hard that no one will be able to figure them out.

Students should be clear about who would fit their rule *and* who would not fit; this eliminates rules like WEARING DIFFERENT COLORS, which everyone will probably fit. It's also important to pick a rule about something people can observe. One rule for classifying might be LIKES BASEBALL, but no one will be able to guess this rule by just looking.

Guess My Rule can be dramatic. Keep the mystery and drama high with remarks such as, "That was an important clue," "This is very tricky," "I think Jamal has a good idea now," and "I bet I know what Annie's theory is."

It is surprising how hard it can be to guess what seems to be an obvious rule (like WEARING GREEN). It is often difficult to predict which rules will be difficult. Sometimes a rule you think will be tough is guessed right away; other times, a rule that seemed obvious will turn out to be impossible.

Give additional clues when students are truly stuck. For example, one teacher chose WEARING BUTTONS as the rule. All students had been placed in one of the two groups, but still no one could guess. So the teacher moved among the students, drawing attention to each in turn: "Look carefully at Dominic's front. Now I'm going to turn Seung around to the back, like this—see what you can see. Look along Maya's arms." Finally, students guessed the rule.

Classification is a process used in many disciplines, and you can easily adapt Guess My Rule to other subject areas. Animals, states, historical figures, geometric shapes, and types of food can all be classified in different ways.

Sketching Data

Graphs are traditionally the focus for instruction in statistics in the elementary grades, even though the process of doing statistics involves much more than making and reading graphs. Most of us think of graphs as the endpoint of the data analysis process. Mathematicians and scientists, however, use pictures and graphs during the process as tools for understanding the data and seeing overall characteristics more clearly.

Our students must gain facility with creating pictures, diagrams, tallies, tables, graphs, and concrete models that provide a first quick look at data. We call these representations "sketches" to emphasize that they are working representations rather than final products. Sketches should be easy to make and easy to read; they should not challenge students' patience or motor skills. Sketches:

- can be made rapidly
- reveal important features of the data
- are clear, but not necessarily neat
- don't require labels or titles (as long as students are clear about what they are looking at)
- don't require time-consuming attention to color or design
- may not be precise, straight, or perfectly aligned

Sketch graphs can be made with pencil and paper, interlocking cubes, and stick-on notes. Cubes and stick-on notes offer flexibility since they can easily be rearranged. Encourage students to construct concrete and pictorial representations of their data using interlocking cubes, pictures, or even the actual objects. Model for students a variety of quick sketches, including forms of simple bar graphs and tallies.

Encourage students to invent and use different forms until they discover some that work well in organizing their data. Students in the third grade are capable of inventing simple—but effective—sketches and pictures of the data they collect, like the examples shown below and on the preceding pages.

At this age level, stress making clear and organized representations, but do not require precision. Students may not make uniform squares for their bar graph or line up their pictorial symbols evenly, but if their sketch helps them describe and interpret their data, they are well on their way to developing important data analysis skills.

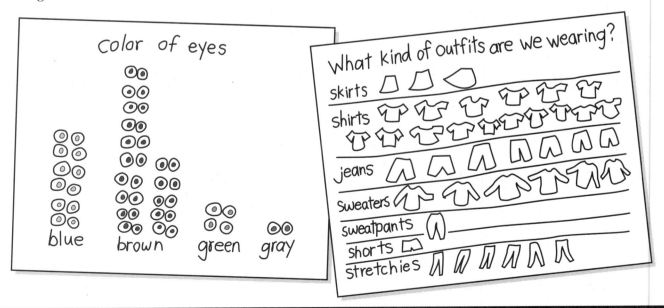

Playing Guess My Rule

This class is playing Guess My Rule (p. 48), and the teacher's secret rule is WEARING GREEN.

Su-Mei and Mark both fit the rule I'm thinking of. Let's have people who fit my rule stand here. [*Su-Mei and Mark stand up in front of the chalkboard.*] **Who thinks they know someone else who might fit this group? Don't guess the rule; just tell me another person you think might fit.**

Liliana: Do I fit?

Yes, you do fit my rule. [*Liliana joins Su-Mei and Mark. It happens that all three children have black hair. This characteristic is visually striking when they all stand together.*]

Maria [*who has black hair*]: I think I fit.

No, you don't fit the secret rule, but I bet I know what you were thinking about. Stand over by my desk to start the "people who don't fit the rule" group. Maria is an important clue.

Cesar: I know what the rule is! It's …

Don't say the rule yet. If you think you know, tell me someone else who fits.

Cesar: Um … [*Looks around, can't find anyone.*]

What about yourself?

Cesar: I don't think I fit.

OK. Go stand with Maria so people have more clues for who *doesn't* fit.

[*Later*]

Su-Mei: Does Christina fit? [*Christina is wearing a green shirt and pants.*]

Yes, she does fit—that's another important clue.

Samir: I know! I know!

Others: Me, too! I know the rule!

OK, let's see if anyone else fits the rule. Then you can say what you think it is.

In this conversation, the teacher keeps the focus on looking carefully at all the evidence, rather than on getting the right answer quickly. She uses Maria's sensible guess to point out the value of negative information: Even though Maria does not fit the rule, she provides an important clue in narrowing down the possibilities. By prolonging the discussion and gathering more clues, the teacher gives more students time to think and reach their own conclusions.

Handfuls of Cubes and Other Objects

What Happens

Students determine how many objects—such as beans, cubes, or tiles—they can hold in each of their hands. They count the objects in two different ways. They combine the amounts they can hold in their hands and then compare the amounts they can hold in their right and left hands. They discuss and record their strategies for combining and comparing amounts using numbers, words, and sketches. Their work focuses on:

- using grouping to count
- developing strategies for combining and comparing amounts
- recording strategies for counting, combining, and comparing

 Ten-Minute Math: Exploring Data Once or twice during the next few days, try the activity Exploring Data. This activity is designed to be done during any spare ten minutes you have outside of math class, to give students continued opportunities to collect, graph, and describe real data.

Choose a question that involves data students know or can observe. For example: How many buttons do you have today? What month is your birthday? What did you have for dinner last night? Are you wearing shoes with laces? How did you get to school today?

Quickly collect and graph the data using a line plot or bar graph. Ask students to describe the data.

What do you notice? Where do most of the data seem to fall? What seems typical or usual for this class?

Also ask students to interpret and predict:

Why do you think that the data came out this way? Does anything about the data surprise you? Do you think we'd get similar data if we collected it again tomorrow? Next week? In another class? With adults?

List any new questions that arise from considering these data. For more information about this activity, see p. 89.

Materials

- Interlocking cubes: 2 open containers
- Two or more open containers of other items to grab (kidney or lima beans, square tiles, pennies, buttons)
- Calculators (1 per student or pair)
- 100 charts
- Student Sheet 16 (3–4 per student)
- Student Sheet 17 (1 per student, homework)
- Overhead projector

Handfuls of Cubes

Show students an open container of interlocking cubes.

How many cubes do you think you can grab with one hand from this container?

Get a few estimates. Then demonstrate how to grab the cubes and have a volunteer come up and grab a handful.

At this point you might want to discuss guidelines for grabbing cubes. For example, should you be allowed to scoop up the cubes so they are piled high on your hand? Should you count dropped cubes? As questions arise, allow students to establish appropriate rules. You might want to have volunteers demonstrate how to grab cubes according to the class rules.

Today and tomorrow you will be grabbing handfuls of objects like cubes, beans, and tiles. You will be comparing how many you can hold in your right hand with how many you can hold in your left hand. Which hand do you think can hold more—your right or your left? Would they hold about the same?

Ask students to explain the reasons for their predictions.

After you determine how many cubes you can hold in each hand, you will compare the number of cubes in your left hand with the number of cubes in your right hand. You'll figure out which hand held more cubes and how many more.

Counting Handfuls   Grab a handful of cubes and have a student count how many you grabbed. Emphasize the idea of double-checking.

It's always a good idea to double-check your work. Chantelle just counted the cubes by 2's and found that there were 16. Does anyone have another way to count them?

Ask for a volunteer to double-check by counting the cubes in a different way. Record the hand (right or left) and the number of cubes you could hold, on the board or overhead.

Now I want to see how many cubes I can hold in my [left] hand. Before I grab, how many cubes do you think I can hold?

After students have volunteered estimates, repeat the procedure for grabbing and counting using your other hand, and record this amount. Again, emphasize the importance of counting the cubes in two different ways as a way of double-checking.

When students count the cubes during the handfuls activity, they will sometimes come up with different numbers. Discuss what to do when this happens, even if it doesn't happen in the demonstration.

Sometimes when you double-check an answer, you will come up with two different numbers. What might you do if this happens?

Students then generate strategies for dealing with this situation. They typically suggest recounting or asking a friend to count.

Comparing Handfuls   Ask students to look at the handful data you have recorded:

LEFT HAND 16   RIGHT HAND 21

How would we figure out how many more cubes I could hold in one hand than in the other hand?

Have the class share some strategies for finding the difference between these two numbers. Emphasize that figuring out the solution in more than one way is a way of double-checking.

Student strategies may vary. When the numbers are small or close in amount, the answer will seem obvious. It is important to have students focus on their *strategy* for determining the difference. Do they count on from the smaller number? Do they count backward from the larger number? Do they use a tens number as a landmark? ("From 16, it's 4 to get to 20, then add 1 more.") Help students articulate their strategies with familiar numbers so they will be able to apply these same strategies when the numbers are larger or farther apart.

Combining Handfuls   Now ask students to look at the handful data in another way:

What if I wanted to combine the amounts in my two handfuls of cubes? How could I do that?

Have students share ideas for adding the number of cubes in your left hand to the number in your right hand. Take note of the addition strategies students are using. See the **Teacher Note,** Two Powerful Addition Strategies (p. 30).

Some children may suggest counting on from one number; some might suggest pushing all the cubes together and recounting them; others might ignore the cubes and focus on the numbers. In the last case, ask yourself: Are they using the tens as helpful numbers—adding them first, then adding the ones? The **Dialogue Box,** Strategies for Combining Handfuls (p. 67), will give you some ideas of strategies your third graders might use.

Grabbing Handfuls

Explain that for the remainder of this session and the next, students will be counting, comparing, and combining handfuls of different objects. In addition to the cubes, show students the containers of beans and other objects you have chosen. If possible, you might want to put one container on every table or group of desks. Or, establish a rule, such as, "Only four people can be at any container at a time."

Distribute several copies of Student Sheet 16, Handfuls, to each student. Students will use a separate sheet to record every item they grab, the data for their right hand and left hand, the comparison of handfuls, and the combination of handfuls. Point out that students will need to explain how they figured out the comparison and combination by using words, pictures, and numbers.

Explain that everyone needs to collect data about *cubes,* because they will use these data for tomorrow's session. In addition, they can collect data about any of the other objects. Each time they go to a different container, they will need to fill out a new sheet. Suggest that they might want to collect data on the same object more than once.

Calculators should be available to students during this activity. If they choose to use the calculator for combining or comparing amounts, suggest that they also find a way to double-check the calculator answer and record this on their sheet. Encourage them to work with partners and to use each other as resources if they have questions. Some students may want to use the 100 chart for comparisons. See the **Teacher Note,** Using Concrete Materials for Addition and Subtraction (p. 65), for more details.

Allow the remainder of this session and half of the next session for the Handfuls activity. During this time, observe each group to get a sense of how they are counting their handfuls and the strategies they are using for combining and comparing.

Note that there is a homework activity for students to complete after Session 3 (see p. 64), asking them to collect more data on Student Sheet 17, Handfuls at Home. The completed homework will be used in class discussion at the end of Session 4.

Looking at Data from Handfuls

As students gather their handful data about cubes, you might have them report it to you for recording on the board. Record the number of cubes each student held in one hand (either right or left; choose one). List the numbers without names to discourage competition about who held the most. At this point, don't put the numbers in order. Alternatively, you might want to have students record their totals on the board sometime during the first half of Session 4.

Using a Line Plot   About halfway through Session 4, bring the class together to discuss the cube data. Begin the discussion by asking students what they notice about the number of cubes in a handful. After they make a few observations, introduce the line plot as a way to look at the data. See the **Teacher Note,** Line Plots: A Quick Way to Show Data (p. 66).

I'm going to put these numbers on a graph called a *line plot*. A line plot is like a number line. What's the smallest handful we have? What's the largest handful we have?

Draw a line plot on the board or overhead. Use the smallest and largest count in a handful to establish the range on your line plot. Add at least one number below the smallest handful and one number above the largest to show that there are other amounts that could be held. Put the student data on the plot. Students can help by reading the numbers listed on the board aloud to you. After all data have been recorded, ask students to look at the graph:

Looking at the graph, what do you notice about our handful data?

Students might comment on the frequency of certain numbers ("5 people held 9 cubes"), the number of cubes grabbed by the greatest number of people ("11 is the most common"), or what the graph looks like ("It looks like skyscrapers").

You might ask students to compare certain pieces of data:

Compare the number of people who held 10 cubes to the number of people who held 11 cubes.

Compare the number of people who held more than 13 cubes to the number of people who held less than 13 cubes.

Handfuls: What Do You Think?

End Session 4 by posing the following question:

Consider for a moment all the data you have collected about handfuls of different objects, including the handfuls you did for homework. What comments or observations can you make about your handfuls?

Some students may compare the amounts of objects they could hold. ("I held more beans than tiles.") Others may generalize about the how the size of an object is related to the number that can be held. ("The bigger the object, the less you can hold.") Still other students might comment on the activity itself. ("It was fun." "Some things were easy, like the cubes, but others were hard.") By soliciting feedback, you not only get a sense of how students are experiencing the curriculum, but also gather information about how you might extend, adjust, or alter activities.

❖ **Tip for the Linguistically Diverse Classroom** Encourage students to demonstrate their observations. If they are able, ask them to verbalize what each demonstration shows. For example, students might show that they can hold more beans than tiles. Or, they might show that the bigger the object is, the fewer the number that will fit in their hand.

Students will resume their work with comparing and combining handfuls in the grade 3 Addition and Subtraction unit, *Combining and Comparing*. If you feel your students could benefit from or want to spend more time collecting handful data, consider making the materials available before school or during free periods.

Sessions 3 and 4 Follow-Up

 **Homework**

Handfuls at Home After Session 3, send home Student Sheet 17, Handfuls at Home. Possibilities for handfuls at home include marbles, pennies, building blocks, letter tiles from games, buttons, plastic foam packing pieces, or popcorn (popped—unpopped kernels are too small). Students collect data for their own handfuls and data from one other person at home. The information they collect can be discussed as part of the final activity.

 **Extensions**

Handfuls from Other Students If your students need more experience with counting, combining, and comparing handfuls, you might have them collect handful data from a class of younger students.

Handfuls from Adults For an interesting comparison with their class data, students could collect handful data from adults in school or adult family members who stop in at the beginning or end of the day.

Using Concrete Materials for Addition and Subtraction

Many of your students will need to model problems in a way that helps them see the problem situation and what they are trying to find out more clearly. Help students use cubes, counters, or 100 charts to show you how they are thinking about the problem. Here are some examples of how students can use these materials:

Cubes or Counters Suppose a student is comparing the 34 beans grabbed with the right hand to the 26 beans grabbed with the left hand. The student might build a tower of 34 cubes and a tower of 26 cubes and then compare the two by laying them on their sides, one on top of the other. Then the students can count how many more cubes are in the longer tower.

Another student might count on from 26 by building a tower of 26, adding cubes in a different color up to 34, and counting the newly added cubes to get the difference. A third student might see this problem as finding the difference between 34 and 26, and build 34 with the cubes, take 26 away, and count how many are left. (It is not likely, however, that many students will see a problem like this as a subtraction problem.)

As students are counting cubes, gradually help them move away from counting by 1's. Some students cling to counting by 1's because they feel it is the only way they can be sure of their accuracy. However, counting by 1's is difficult, inefficient, and prone to error—for anyone. Help students build their confidence and skill in counting by 2's, 5's, and 10's. If you keep interlocking cubes stored in towers of ten, students will be able to work with them more easily.

100 Chart On the 100 chart, a student can place a counter or a mark on 26 and then find how many steps or jumps it takes to reach 34. At first, students may count by 1's; for this particular example, where the difference is less than 10, that is an appropriate strategy.

When students are comparing two numbers that are farther apart, such as 23 and 69, encourage them to figure out how to make bigger jumps. One student might jump from 23 to 30, then to 40, 50, 60, and finally to 69. You might urge this student to jot down the intermediate differences as jumps are made along the chart (7, 10, 10, 10, 9). Another student might start at 23 and jump to 33, 43, 53, 63, and then to 69, counting the jumps as "10, 20, 30, 40 plus 6, that's 46" while moving the counter on the 100 chart.

One of the biggest problems students have on the 100 chart is understanding where to start counting. When finding the difference between 26 and 34, for example, many students count both 26 and 34, ending with a difference of 9 instead of 8. Help these students by working with much smaller numbers:

What if you only grabbed 5 cubes in one hand and 7 in the other? Why don't you say 7 is 3 bigger than 5?

Students who persist with counting the beginning number should return to using cubes or counters. Reformulate the question in this way:

Here are the 26 beans you grabbed with your left hand. I'll make them with red cubes. I want you to add on blue cubes until you have 34. Then we can see how many extra beans you'd need to get the 34 beans you grabbed with your right hand.

A line plot is a quick way to organize numerical data. It clearly shows the range of the data and how the data are distributed over that range. Line plots work especially well for numerical data with a small range, as would be typical for third grade handfuls of the same object.

A line plot is most often used as a working graph and is especially useful as an initial organizing tool for work with a data set. It need not include a title, labels, or a vertical axis. A line plot is simply a sketch showing the values of the data along a horizontal axis and X's to mark the frequency of those values. For example, if 15 students have just collected data on the number of tiles they could hold in their right hands, the data could be represented in a line plot like the one shown below.

From this display, we can quickly see that the range is from 7 to 14, and that most students grabbed between 12 and 14 tiles.

One advantage of a line plot is that we can record each piece of data as we collect it. To set up a line plot, start with an initial guess from students about what the range of the data is likely to be:

What do you think the lowest number should be? How high should we go?

Leave some room on each end of the line plot so you can lengthen the line later if the range includes lower or higher values than you expected.

By quickly sketching data in line plots on the board, you provide a model of how such plots can provide a quick, clear picture of the shape of the data.

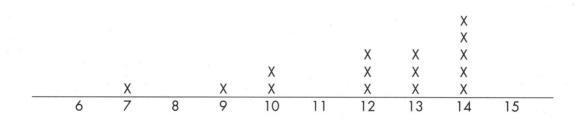

Strategies for Combining Handfuls

During a discussion in the activity Handfuls of Cubes (p. 60), the students in this class are sharing strategies for *combining* handfuls of cubes. They are using the handful data from their teacher:

Left hand   16
Right hand   21

What if I wanted to combine my handfuls of cubes? How could I do that?

Yvonne: You could just take that pile of cubes from your left hand and that pile from your right and moosh them together and then count how many cubes there are in the big pile.

How could you write that down so that some-one who wasn't here could understand what you did?

Yvonne: Well, I guess you could use words and just write, "I mooshed them together and then counted."

How did you count?

Yvonne: From 1. You know, 1, 2, 3, 4, 5, all the way until there were no cubes left.

It would be important to say that part too.

Rashad: It's a little hard just by looking at it, but I think I might say: 10 plus 20, that's 30, then 6 plus 1 is 7, so 30 plus 7 is 37.

So you would think about how many groups of ten you had and add them first.

Rashad: Right. Tens first.

Saloni: Well, I would say 16, then I would count up that pile of 21.

How would you count up the pile of 21?

Saloni: Can I show it? See I would go 16 [*gesturing to the pile of 16*], 17, 18, 19, 20, 21, 22, 23, 24, ..., 35, 36, 37. [*She moves one cube away from the pile of 21 as she says a number.*]

Does anyone have any other ideas for combining these two groups of cubes?

Aaron: You could use a calculator—then it would be easy! First you punch in 16, and then you punch in 21, then the equals button. And that would tell you how much.

Yes, you could use a calculator to solve this problem. If you decide to use a calculator, you should also double-check the calculator by using a different strategy for combining your handfuls. Any more ideas?

Yoshi: Well it's sort of like Rashad's way. I'd say 20 plus 16—that's 36—then add one more from the 21.

How would you write that down to show some-one what you did?

Yoshi [*goes to the board and writes*]:

$$20 + 16 = 36$$
$$36 + 1 = 37$$

Exploring Odds and Evens

What Happens

Session 1: Adding Odds and Evens   Students refer to previous experiences in this unit, such as their list of doubles, to discuss what they already know about odd and even numbers and to suggest ideas for a class chart. They explore what happens when they add two even numbers, two odd numbers, or an odd and an even number by building addition combinations with cubes.

Session 2: Odds and Evens on the Calculator Students explore the calculator as a tool for mathematics. As they make up and solve problems on the calculator, you have a chance to see how comfortably they use this tool. A class discussion helps students develop an awareness of the decimal point and what it means. Students do simple division problems on the calculator to help them understand the meaning of 0.5. Working in pairs, students find numbers of apples that can be evenly divided between two people, with and without using halves.

Session 3: What We've Learned About Odds and Evens   Students discuss as a class what they discovered about odd and even numbers and add new ideas to the list (or change old ideas to reflect new understandings). As an assessment activity, they choose and write about one of the ideas, giving reasons and examples to support the idea.

Mathematical Emphasis

- Exploring the characteristics of odd and even numbers
- Examining how odd and even numbers behave when they are combined
- Using evidence gathered from examples to make conjectures about the ways numbers behave
- Continuing to develop familiarity with addition combinations
- Becoming familiar with the calculator as a tool for mathematics
- Working with wholes and halves
- Developing awareness of the decimal point and its meaning
- Communicating orally and in writing about mathematical ideas

IDEAS ABOUT ODD AND EVEN.

▶ Even numbers can be split into two equal parts. Odds can't.

▶ All the even numbers are in the twos table. No odds are in the twos.

What to Plan Ahead of Time

Materials

- Students' sets of Addition Cards saved from Investigation 2 (Session 1)
- Interlocking cubes: 60 per pair (Sessions 1–2)
- 100 charts (Session 2)
- Calculators: at least 1 per pair (Session 2)
- Chart paper (Sessions 1–3)

Other Preparation

- Make sure the classroom list of doubles remains posted where students can see it.
- Duplicate student sheets (located at the end of this unit) in the following quantities. If you have Student Activity Booklets, no copying is needed.

For Session 1

Student Sheet 18, Is the Sum Odd or Even? (p. 122): 1 per pair or group

For Session 2

Student Sheet 19, Dividing Apples: No Halves! (p. 123): 1 per pair

Student Sheet 20, Dividing Apples: Using Halves (p. 124): 1 per pair

Student Sheet 21, Odd and Even Numbers (p. 125): 1 per student (homework)

For Session 3

Student Sheet 22, Writing About Odd and Even Numbers (p. 126): 1 per student

Adding Odds and Evens

Materials

- Interlocking cubes (60 per pair)
- Chart paper
- Students' Addition Cards
- Student Sheet 18 (1 per pair or group)

What Happens

Students refer to previous experiences in this unit, such as their list of doubles, to discuss what they already know about odd and even numbers and to suggest ideas for a class chart. They explore what happens when they add two even numbers, two odd numbers, or an odd and an even number by building addition combinations with cubes. Their work focuses on:

- thinking about characteristics of odd and even numbers
- experimenting with combining odd and even numbers
- developing conjectures about how odd and even numbers behave

 Ten-Minute Math: Exploring Data   Once or twice during the next few days, continue to do the Exploring Data activity during any spare ten minutes you have outside of math class. You may want to collect data on different days for the same question; for example, How many pockets do you have? Or, how many buttons do you have?

Make a separate line plot for each day's data, and save the graphs so students can think about how the data change over a few days. For more information on this activity, see p. 89.

Activity

What Is an Even Number?

In this investigation, students will be exploring the behavior and characteristics of odd and even numbers. Much of the work they did in Investigation 2 with doubles and halves is relevant to the work in this investigation. Encourage students to look back at their work from Investigation 2 and to contribute ideas based on that work as they discuss odd and even numbers.

As you look at the suggestions for how to begin this discussion, think about ways you might modify your approach, depending on what your students have already noticed about odd and even numbers.

You might start by drawing attention to the list of doubles that the class made in Investigation 2. Ask students again what they notice about the sums of the doubles, or refer to the list of observations you made with them. If someone has noticed that all the sums are even, focus on this observation (or introduce the idea yourself):

You noticed before that sums of the doubles are all even numbers (or, I notice that all the sums of the doubles are even numbers). Do you agree? What is an even number?

Try to probe students' understanding of what *even numbers* are. This discussion should last at least 5 minutes, but it may take longer. See the **Dialogue Box,** What Do You Mean by Even? (p. 73), for an excerpt from the discussion in one classroom.

On chart paper, begin a list of students' ideas about odd and even. This list might include statements like these: "It's a number that doesn't have a middle." "It's the numbers that end in 0, 2, 4, 6, and 8." The Dialogue Box just mentioned gives other examples of observations students have made. Include ideas students have that still need to be tested (and might be found wrong). Some teachers list conjectures—statements for which you have good evidence but have not yet been proven—along with the name of the student who first had the idea:

> Yvonne's Conjecture: Even numbers don't have a middle.

When the discussion has gone as far as it can for now, tell students they will be finding out more about even and odd numbers during this class and the next one, when they will continue the discussion and add any new discoveries to the list. Add the title *Ideas About Odds and Evens* to the chart and post it in the classroom.

Which Are Odd? Which Are Even?

Each pair (or small group) needs one set of Addition Cards (saved from Investigation 2), 60 interlocking cubes, and a copy of Student Sheet 18, Is the Sum Odd or Even?

We're going to find out more about what happens when you add even numbers or odd numbers. You've already said that when you add a number to itself—the doubles—it always comes out even. I'd like you to think about more of your addition combinations and see when addition comes out even and when it comes out odd. Let's try 8 + 6.

Ask all students to build 8 and 6 separately with interlocking cubes, even if they know this addition combination.

Is 8 odd or even? Is 6 odd or even? How do you know? When you add 8 and 6, is the sum odd or even? Why do you think it's even?

Pairs (or groups) choose ten Addition Cards from their set. They build each combination with cubes, decide whether each number is odd or even, decide whether the sum is odd or even, and record their results on Student Sheet 18. Students build the two numbers with cubes even if they already know the addition combination; the cubes help them observe the structure of odd and even numbers and what happens when they are combined.

As you watch students work, talk with them about what they are finding out:

What happens when you add two even numbers? Two odd numbers? An odd number and an even number?

Make sure they write anything they notice on the back of Student Sheet 18 so they can bring up their ideas during the class discussion in Session 3.

Session 1 Follow-Up

Extension

Subtraction with Odds and Evens   Some students may enjoy exploring what happens when you subtract an even number from an even number (12 – 6), an odd number from an odd number (9 – 5), an odd number from an even number (12 – 3), or an even number from an odd number (9 – 6). Students who are interested might work on this at home or in school and add their conjectures to the list the class will make in Session 3.

DIALOGUE BOX

What Do You Mean by Even?

In this discussion that occurred during the first session of Investigation 4, these students are thinking and talking about the question, What is an even number?

Kate: Odd is if you have three apples. You couldn't split them with a friend.

Split them how? Why can't I get two and you get one?

Kate: No, it's not even. If we had four, you could have two and I could have two.

Khanh: If you had five apples, we could each get two and split the other in half.

Tamara: If you took pencils or pens and gave an odd number of them to someone, and then you took one away or added one, you'd get an even number.

Sean: Let's say your cousin wanted a watch like yours. You already had one. If you get one for her, it's even.

Ly Dinh, what's your idea?

Ly Dinh: You skip the first number, say the second number, and then you skip the next one.

Elena: You skip every other number.

Does it matter where you start counting?

Ly Dinh: Start at 2, and you can put things in 2 groups that are even or equal.

Elena: It's all the numbers like 2, 4, 6, 8.

Khanh: I'd say it's a number that either had to be 2, 4, 8, 0 or end in those numbers.

Elena: Or 6.

Yvonne: It's a number that doesn't have a middle.

What do you mean, it doesn't have a middle?

Yvonne: It's like, if you take 3, it has a middle, but 4 splits in half and it doesn't have a middle.

I'm not quite sure I get it. Can you show us what you mean?

Yvonne [*Illustrating on the board*]: See, I make 3 like this, and I can circle the middle, but if I make 4, there's no middle, so it's even.

Do the rest of you think that's true? Do even numbers ever have a middle?

Odds and Evens on the Calculator

Materials

- Calculators (at least 1 per pair)
- Interlocking cubes
- 100 charts
- Student Sheet 19 (1 per pair)
- Student Sheet 20 (1 per pair)
- Student Sheet 21 (1 per student, homework)
- Chart paper

What Happens

Students explore the calculator as a tool for mathematics. As they make up and solve problems on the calculator, you have a chance to see how comfortably they use this tool. A class discussion helps students develop an awareness of the decimal point and what it means. Students do simple division problems on the calculator to help them understand the meaning of 0.5. Working in pairs, students find numbers of apples that can be evenly divided between two people, with and without using halves. Their work focuses on:

- becoming comfortable with the calculator to solve computation problems
- recognizing the decimal point on the calculator screen and keyboard
- looking at different ways to write 1/2
- understanding the meaning of 0.5
- dividing even and odd quantities in half

Teacher Checkpoint

Exploring the Calculator

If students have not used calculators during mathematics class before, they will at first be distracted by having them. As with any new material, students need time to explore the calculator and find out what it can do.

Distribute calculators, at least one to each pair, and give students 15 minutes or so for the following task:

With your partner, make up one addition problem and one subtraction problem. Solve the problems on the calculator, then figure out a way to check the calculator by doing the problems in your head or with cubes. I'll come around and help if you can't get the calculator to do the problems correctly.

As you circulate, make sure students know how to clear the calculator between problems and how to use the +, –, and = keys. If they are comfortable with the + and – keys, ask them if they can make up a problem using the × or ÷ keys, or ask them to try some of the doubles or halves problems they did in Investigation 2.

Get a sense of how comfortable students are with the calculator:

- Can they do straightforward computation easily?
- Are they familiar with the symbols on the keyboard?
- Can they read the screen?

If they seem unfamiliar with the calculator, you may need to schedule more sessions like this one or work with individual students. Rather than teaching use of the calculator to the entire class at once, work with small groups or ask students who know how to use the calculator to work with those who don't.

Also, look for examples of students' encounters with the decimal point to bring up during the class discussion, What's the Little Dot?

Discussion: What's the Little Dot?

After students have explored the calculator, hold a class discussion about the decimal point. The decimal point is important: On many calculators, it always appears on the screen to the right of whole numbers. Also, students are likely to see it with fractional amounts when they are just fooling around or experimenting with the calculators, or when they have made an error.

Sometimes students don't see the decimal point (for example, they read 3.25 as "three hundred twenty-five"); this is either because they are ignoring a symbol that has no meaning for them, or because the decimal point is often hard to see on calculator screens. At other times, students see a decimal number as a mistake: "Look, I just divided 98 by 3 and I got this weird number." It's critical that third grade students know how to interpret the decimal point in its broadest sense—that it indicates a small part of a number, less than one—and that they begin to understand the meaning of the decimal 0.5.

The issue of the decimal point likely arose during the 15 minutes while students were exploring the calculators. If you saw one of these occurrences, refer to it to start the discussion. Otherwise, start the discussion by using the apple-halving problems described below.

Annie and Tyrell got this number [*write 432.567 on the board*] **on their calculator. What do you think this little dot means?**

Students will have a variety of ideas: "It's like a comma," "It's like the cents part in dollars and cents," "It means hundreds."

This dot has an important meaning. We're going to find out what it is, but first I want you to try something.

Dividing Apples   Explain that you're going to give students a real situation in which they'll find the dot on the calculator to be useful.

First, suppose we were going to divide four apples evenly between Jennifer and Jamal. What would we do?

Get student ideas, and draw a picture on the board to show what would happen to the apples.

Now let's try it on the calculator. Who knows how to divide 4 into 2 parts on the calculator?

Help students figure out how to do this problem on the calculator by using the keys 4 ÷ 2 = . You can help them interpret this problem as "4 apples divided into 2 groups; how many in each group?" Make sure all students are able to do this problem on the calculator. You may want to try one or two more problems that divide evenly, such as 6 apples split among 2 people and 10 apples split among 2 people.

After students have done this problem successfully on the calculator, ask them what would happen if they had three apples to split between Jennifer and Jamal.

So you think Jennifer and Jamal should each get one and a half apples. Who knows a way to show "one and a half"?

On the board, record ways students have to write "one and a half." Correct any that are incorrect (such as 1²⁄1), but include any reasonable ways (for example, "one apple and half an apple," a picture of a whole circle and a half circle, or "one and ½"). If students don't suggest the fraction notation 1½, write it on the board and explain that it is another way to write one and a half.

What Does 0.5 Mean?   Now ask students to use the calculator to solve the same problem.

Three apples split between 2 people would be set up this way on the calculator: 3 ÷ 2 = . Try it. What does it say on your calculator screen?

Write 1.5 on the board, and ask students to discuss what it means. Make sure they see the decimal point on their screen. After hearing some of the students' ideas, say:

1.5 is another way to write "one and a half." The "point 5" part means a half.

Let's think about another problem. Don't do this one on your calculators yet, because I want you to tell me what you think might happen. If I just have one apple and I want to split it between Dominic and Jeremy, how much will each person get? What do you think the calculator screen will show?

After students predict the answer, have them try it on the calculator. Record what they see on their screen, and discuss what they think it means. They may be surprised that the calculator screen shows 0.5 instead of just .5. You can tell them that this is the way the calculator shows "no whole things (apples) and a half." Conclude the discussion by saying something like this:

When you see a number, then the decimal point, and then a 5, read it "point 5," like 23.5 (twenty-three point five). The "point 5" part always means a half, so 23.5 means 23 and a half. The decimal point always means that there is a part of the number less than one, just like one-half is less than one.

So if you see a number like 3.4455, you know that you have 3 and some small pieces; those small pieces aren't enough to make another whole piece. It's like 3 whole apples, and some pieces of another apple, but not another whole apple.

You may want to try a few more examples, but don't expect all students to understand the meaning of the decimal point. This discussion can only be the beginning of decimal point awareness.

As the decimal point comes up during calculator use throughout the year, help students see the point and think about the two parts of the number it is in: the whole number part, and the "extra pieces." It's not necessary for students to know how to compare these extra pieces—for example, that 0.4 is less than 0.5; they will learn to do this kind of comparison in later grades. For now, 0.5 is the important decimal landmark they will learn.

See the **Teacher Note,** Using the Calculator Sensibly in the Classroom (p. 80), for more tips on integrating the calculator into your mathematics curriculum.

Splitting Numbers in Half

Students work in pairs using cubes, calculators, and one copy of Student Sheet 19, Dividing Apples: No Halves! They write down ten amounts of apples they can split between two people equally *without* using halves, and—without using the calculator—record how many apples each person would get. Students should be encouraged to use the cubes if they need to be sure a particular amount will divide into two even piles.

Pairs then do the same problems on the calculator and write down those results. Emphasize that they should write down their own answer to each problem before they try it on the calculator.

When a pair has found ten amounts they can split equally without using halves, give them Student Sheet 20, Dividing Apples: Using Halves. They repeat the activity, this time finding ten amounts they can split equally only by using halves. Since students may have no previous experience with fractions, allow them to write results like "three and a half" in their own way.

As you talk with students, ask them to write down any general ideas they have about which numbers split evenly, without halves, and which don't. Encourage students to experiment.

For example, if the first five amounts a pair has written on the first sheet are 2, 4, 6, 8, and 10, discuss the pattern they are using and whether it would continue, but then ask if there are higher numbers they want to test rather than continuing with 12, 14, and 16. They can write about their pattern on the other side of the paper. Some students may want to test high numbers like 1000, even though 1000 is an unlikely number of apples to split between two people!

Before the end of the session, be sure students know how and where to store the calculators. See the **Teacher Note,** Using the Calculator Sensibly in the Classroom (p. 80), for hints about calculator storage.

Odd and Even Numbers In preparation for the discussion in Session 3 about odd and even numbers, hand out Student Sheet 21, Odd and Even Numbers. Students will use this assignment as a reference during the discussion, and perhaps as a starting point for their assessment task. You might say something like:

Tomorrow we'll be talking about what you've found out about odd and even numbers, and you'll be doing some writing about what you've learned. Tonight, think about at least two things you know now about odd and even numbers. Write them down on this sheet, and bring them to class tomorrow.

❖ **Tip for the Linguistically Diverse Classroom** Students who are not writing comfortably in English will be able to convey many thoughts simply by using numbers, drawings, and the words *even* and *odd*. For example:

2, 4, 6, 8, 10, 12, 14 = even

1, 3, 5, 7, 9, 11, 13 = odd

$$\left.\begin{array}{l} 4 + 4 = 8 \\ 6 + 6 = 12 \\ 2 + 2 = 4 \end{array}\right\} \text{ even}$$

even   odd

 represents the "even" drawing and the "odd" drawing with arrow.

Using the Calculator Sensibly in the Classroom

Why Use Calculators in the Elementary Classroom? Many people are still opposed to the use of calculators in the elementary classroom. You may get questions from parents or other educators, or you may hear statements such as: "Students should not use calculators until they know the arithmetic facts and procedures," or "Students should use calculators to check only after they have solved a problem."

By keeping calculators out of the hands of young students, while they see adults all around them using them, we communicate to students that doing mathematics in school is nothing like doing mathematics outside of school. It is as if all of us used clocks at home, but insisted that students tell time by the sun while in school. Calculators have a critical role in our society. Increasingly sophisticated calculators are being developed and used in settings ranging from high school mathematics courses to jobs in science, business, and construction. Students must learn how to use the calculator appropriately, just as they must learn to read a clock, interpret a map, measure with a ruler, and use coins.

At third grade, students should:

- recognize, interpret, and use symbols on the calculator keys to do addition, subtraction, multiplication, and division

- use the constant function of the calculator to skip count (this activity is introduced in the grade 3 unit, *Landmarks in the Hundreds*)

- recognize the decimal point when they see it on the calculator screen and keyboard and begin to understand its meaning

When Should Students Use Calculators?
Students should always have access to calculators as a tool for doing mathematics, just as they have access to 100 charts, pattern blocks, and interlocking cubes. Students love to use what they perceive as real, adult tools. When the use of calculators is permitted only occasionally, students become excited and distracted on those occasions and focus on the calculators rather than on the mathematics.

As with any tool, it takes practice to use the calculator efficiently, accurately, and appropriately. To learn when and how calculators are the best tool and when they are not, students must use them frequently in their work. The calculator does not substitute for work with materials or for the development of strong mental arithmetic strategies. You will need to make sure that students get experience with all of these separately, and also have opportunities to choose from among many methods to solve problems.

We discourage thinking of the calculator as a tool to "check" with, as if other methods are somehow more fallible. Any method may be used to check any other. In fact, it's quite easy to make a mistake on a calculator. Throughout this curriculum, we encourage students to solve computation problems in more than one way to double-check their accuracy. If they have solved a problem using one mental arithmetic strategy, they might also solve it using another mental arithmetic strategy. If they have solved it with materials first, they might then solve it with a calculator. If they have solved it with a calculator, they can try solving it mentally.

Access and Storage Students should understand that calculators are an available math material, just like pattern blocks and interlocking cubes. They need to know where they are stored, how to take them out, and how to put them away. If you have a calculator for each student, and if your students can accept responsibility, it is helpful to allow them to keep their calculators with their other materials, at least for the first few weeks of school, so that they get used to having and using them.

In many situations, it is not practical to have students keep calculators. One teacher numbered each calculator, stored them in boxes on a shelf, and assigned a number to each student. This system gave the students a sense of ownership of the calculator as a tool they could use, while also helping the teacher keep track of them. If fewer calculators are available, each could be assigned to a pair or a small group.

What We've Learned About Odds and Evens

What Happens

Students discuss as a class what they discovered about odd and even numbers and add new ideas to the list (or change old ideas to reflect new understandings). As an assessment activity, they choose and write about one of the ideas, citing reasons and examples to support the idea. Their work focuses on:

- using evidence to develop general statements about characteristics of numbers
- organizing a piece of writing about mathematics
- using words, pictures, and numbers to communicate about mathematical ideas clearly
- supporting mathematical statements with arguments and examples

Materials

- Completed Student Sheet 21
- The classroom list, *Ideas About Odds and Evens*
- Student Sheet 22 (1 per student)
- Chart paper

What We Found Out About Odd and Even

In this discussion, students consider evidence from their work during Investigation 2 and state general conclusions they have come to through their explorations. Your job during this discussion is to add more student ideas to the list of conjectures about odd and even numbers, to get other students to comment on these conjectures, to encourage students to bring up examples and counterexamples, and to establish ground rules for this kind of conversation (for example, people respect each other's ideas; people can always revise their thinking when they hear new evidence).

Who has something to say about odd and even numbers? What did you find out when you added them? What did you find out when you divided them on the calculator?

Students can begin by contributing the two ideas they wrote for homework. You may want to list a few student ideas and then return to one or two of them for a fuller discussion:

Khanh says that when you add two even numbers, you always get an even number. Let's think about that one. Who has something to say about this idea?

Saloni said she thought two odd numbers would make an odd number, but she found out that wasn't true. What do you think? What happens when you add two odd numbers?

Laurie Jo found out that when she splits an odd number in half, she always needs to use halves. Do you agree with that? Is that always true?

Spend 10 to 15 minutes on this, or longer if your students are sustaining a good discussion. Challenge students to think about whether their conjectures will always be true and, if so, why. Examples of this type of discussion are given in the **Dialogue Box,** Two Odds Make an Even (p. 85).

This discussion need not lead students to firm conclusions. What's important is that they participate in considering evidence, developing ideas based on that evidence, and testing their ideas by looking at additional examples or counterexamples.

Students will encounter ideas about odd and even numbers frequently. Consider keeping their list of observations posted in the classroom; you may find yourself returning to it throughout the year as these ideas surface in new contexts.

When you add two evens you get an even answer.
8 + 6 = 14
4 + 6 = 10
2 + 2 = 4

When you add odd and even, you get an odd answer.
3 + 2 = 5
7 + 8 = 15
6 + 1 = 7

In the previous two sessions, students have used cubes and calculators to explore some characteristics of odd and even numbers. They have discussed their ideas with a partner and then in a whole-class discussion. In this assessment activity, they will explain in writing one idea or conjecture that they believe to be true about odd and even numbers.

Distribute Student Sheet 22, Writing About Odd and Even Numbers, to each student.

On this sheet, write about one idea you think is true about odd and even numbers. You can choose one of the ideas on our class list, or any other idea you may have. Write why you think this idea is true. Give examples of your idea using words, numbers, and pictures.

Try to explain your idea so that someone who has not studied odd and even numbers could understand what you are saying. As I read what you are writing, I'll probably have questions. I may ask you to add information that will make your writing clearer.

As students are writing, insist that they give reasons for their idea and include examples.

❖ **Tip for the Linguistically Diverse Classroom**   Refer to the suggestions on p. 79 for how students with limited English proficiency can express their thoughts using numbers, simple drawings, and the words *even* and *odd.*

Most students will need to add more ideas to what they first write. As you circulate and read what students are writing, ask questions and suggest what they may need to add. See the **Dialogue Box,** Helping Students Clarify Their Ideas in Writing (p. 86), for an example of an interaction between student and teacher.

About the Assessment in this Unit (p. I-20) offers some suggestions for evaluation. This set of papers can give you a sense not only of what your students know about odd and even numbers, but also of how well they are able to describe their thoughts in writing. Just as students will vary in their mathematical understanding, they will also show varying degrees of comfort and ability in writing about their thinking.

Students will continue to formulate ideas about odd and even numbers throughout the year; likewise, they will have further opportunities to write about their thinking in other *Investigations* units.

Choosing Student Work to Save

As the unit ends, you may want to use one of the following options for creating a record of students' work in this unit.

- Students look back through their folders or notebooks and write about what they learned in this unit, what they remember most, and what was hard or easy for them. You might have students do this work during their writing time.

- Students select one or two pieces of their work as their best work, and you also choose one or two pieces of their work, to be saved in a portfolio for the year. You might include students' solutions to the assessments Writing About Addition (p. 43) and Writing About Odd and Even (p. 83). Students can create a separate page with brief comments describing each piece of work.

- You may want to send a selection of work home for families to see. Students write a cover letter, describing their work in this unit. This work should be returned if you are keeping year-long portfolios.

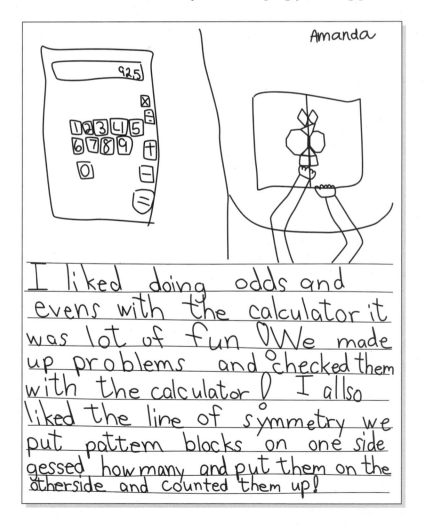

I liked doing odds and evens with the calculator it was lot of fun We made up problems and checked them with the calculator I also liked the line of symmetry we put pattern blocks on one side gessed how many and put them on the otherside and counted them up!

DIALOGUE BOX

Two Odds Make an Even

These students are sharing their ideas during the concluding discussion for this unit, when they are asked to consider what they have learned about odd and even numbers.

Maria: I added 10 and 10, and that's 20, so it's even.

Jeremy: Jamal and I added 6 and 8, and that's 14.

So what did you find out?

Jeremy: When we added two evens, we got an even.

Maria: It has to be even.

Why do you say it has to be even, Maria?

Maria: Because an odd number has to have an extra number, and there isn't an extra number, so it has to be even.

Maria says that if you add an even number and another even number, the sum will be even. What do other people think?

Dylan: That's what we got, too. If the two numbers are even, it's always even.

The sum is even for an even plus an even … but does anyone have an idea why?

Jennifer: Because it can't be odd. There's no way.

Elena: 'Cause look. Here's a 6 and here's an 8 [*shows a cube tower of 6 and a cube tower of 8*]. They're even, right? So if you put them together, they're still even.

What about for two other even numbers. Will the sum always be even?

Jeremy: If it's even, you can break it into twos. So [*breaking the cube towers into pairs*], 2 and 2 and 2 and 2. If you put them together or take them apart, it's still all 2's.

Jennifer: We found out that two odds make an even.

Elena: Yeah, when we added 3 and 9, we get 12. That's even.

Dylan: It's always even, no matter what you do. Everything we got was even.

What happened when you added two odds?

Dylan: It's even. Two evens is even and two odds is even.

So Jennifer and Elena and Dylan are saying that if you add two odds, you get an even. Is that always true? Can anyone explain why that might be true?

Midori: With two odds, it's like, well, we did 7 and 5. See, it's really just like 6 and 6. If you take one off the 7 and put it on the 5, it's all even again [*demonstrates with cubes*].

So that's two odds making an even. Everyone build 7 and 5 with your cubes. Can anyone else add to what Midori says?

Michael: Because it's almost the same as two evens, but they each have an extra, so that's another even, so it's still even.

Yvonne: It's like what Jeremy said about 2's.

Mark: I agree with Michael, because it's like two evens. If you just had 6 and 4, it would be even, but there's one extra on each and that makes another 2, so it's still even altogether.

Helping Students Clarify Their Ideas in Writing

During the final assessment activity for this unit (p. 83), Maya first wrote the following about her ideas about odd and even numbers:

> What I think is that if you add two even numbers you'll always get an even number, like 6 + 6 = 12.

That's a good start. You've said what your first idea is very clearly, and you've given an example. Now you need to say something about why it works that way. Do you think that an even number plus an even number will always give you an even total?

Maya: Um, I'm not sure.

Tell me more about what you think.

Maya: Well, I'm pretty sure. I think it works for all the regular numbers.

Which are the regular numbers?

Maya: Like 2, 4, 6, 8, 10, and all those.

And why do you think it will work for the regular numbers?

Maya: Because there's no extra one, so you can't get an odd number.

What do you mean?

Maya: Like if it's 4 plus 6. Four is even so there's no extra, and 6 is even, so there's no extra, and when you put them together, you can't have an extra, so it's got to be even.

So which numbers don't you think it will work for?

Maya: Um, maybe some really weird numbers, like three billion plus ten thousand.

You need to write down some of what you just told me—so that someone reading your ideas will understand why you think an even number plus an even number will always be even for the regular numbers.

Calendar Math

Basic Activity

Students try to find numerical expressions that are equal to the day's date. For example, if the date is March 19, students look for ways to combine numbers and operations to make 19. Constraints on what numbers or operations they can use push students in developing their arithmetic skills. Discoveries of principles for "expressions that work" become a part of the class's mathematics culture.

Calendar Math is a simple way of providing arithmetic practice and opportunities for students to share mathematical discoveries. Students focus on:

- developing a web of numerical knowledge about any number

- using operations flexibly

- recognizing relationships among operations (for example, that adding and then subtracting the same number has a net effect of 0)

- learning about and using key mathematical ideas, such as the effect of using an operation with 0 or 1

- deriving new numerical expressions by modifying a particular expression systematically (for example, if $2 + 9 = 11$, then so does $3 + 8$, $4 + 7$, $5 + 6$, and so forth)

Materials

Calculators (for variation)

Procedure

Step 1. Pose the problem. For example, "Today's date is September 12. Who can think of a way we could combine numbers to make 12?"

Step 2. List student responses. Their ideas might include expressions like these:

$6 + 6$	4×3	$12 + 0$
$1/2 \times 24$	$3 + (-3) + 6 + 6$	

Step 3. Choose a "favorite expression" for the day. Students choose their favorite from the listed expressions, perhaps the most unusual, or one that uses a new idea. Use the class favorite to write the date on the board: September $(0 \times 12) + 12$.

Variations

Introducing Constraints Introduce constraints based on your class's ease with particular operations and numbers. For example, if students are very comfortable with addition, eliminate addition as a possibility: "Today you can use any operation you want to use, *except* addition." You could also require that a certain operation or kind of number is used. Possible constraints include these:

> You can't use any number that's a multiple of 2.
>
> You can't use addition or subtraction.
>
> You must use more than one operation.
>
> You must use one number that's bigger than 100 (1000, 5000).
>
> You must start with 100.
>
> You must use at least three numbers.
>
> You can't use 0.
>
> You must use at least one number that is smaller than 1.
>
> You must use one negative number.
>
> You can only use 1's, 2's, 3's, and 4's.

Looking for Patterns Encourage students to find expressions that they can alter systematically to find more expressions. Here is a pattern that a student came up with for 20:

> 2×10, $2 \times 9 + 2$, $2 \times 8 + 4$, $2 \times 7 + 6$ …

Developing Class "Rules" Our experience is that, through this activity, new ideas about numbers become part of the culture of the classroom. For example, in one classroom, one student learned about square numbers and the notation for them. Because she used these numbers in Calendar Math, other students became familiar with them and were soon using numbers such as 4^2.

Continued on next page

Other kinds of relationships are often discovered by one student and then become common knowledge. For example, one day when a class was finding expressions equivalent to 14, one student suggested "14 times 0." This remark prompted a discussion of what happens when you multiply a number by 0, and students eventually concluded that the result of multiplying any number by zero is 0. The teacher wrote this on a list of "rules" discovered over the course of the year. Following are some other rules students have discovered during this activity:

A number divided by 1 is the number.

A number multiplied by 1 is the number.

Any number multiplied by 0 is 0.

Any number divided by itself is 1.

Subtract or add 0 to any number and you still have the same number.

Adding lots of 0's doesn't change anything.

You can make any number by adding enough 1's to count up to that number.

Adding a number and then subtracting the same number is like adding 0.

Using the Calculator For a quiet 10 minutes, have students work individually or in pairs on coming up with ways to make the date, using their calculators. Make sure that they record their work on a piece of paper. They can choose their favorite solution and write it on the board. This is a good way for students to explore new keys on the calculator.

Related Homework Options

Planning Ahead. Suggest that students think at home about how they might make the next day's date. Tell students what constraints they are under, and ask them to figure out five different ways to make the date. They can share their favorite in class.

Exploring Data

Basic Activity

You or the students decide on something to observe about themselves. Because this is a Ten-Minute Math activity, the data they collect must be something they already know or can observe easily around them. Once the question is determined, quickly organize the data as students give individual answers to the question. The data can be organized as a line plot, a list, a table, or a bar graph. Then students describe what they can tell from the data, generate some new questions, and, if appropriate, make predictions about what will happen the next time they collect the same data.

Exploring Data is designed to give students many quick opportunities to collect, graph, describe, and interpret data about themselves and the world around them. Students focus on:

- describing important features of the data
- interpreting and posing questions about the data

Procedure

Step 1. Choose a question. Make sure the question involves data that students know or can observe: How many buttons are you wearing today? What month is your birthday? What is the best thing you ate yesterday? Are you wearing shoes or sneakers or sandals? How did you get to school today?

Step 2. Quickly collect and display the data. Use a list, a table, a line plot, or a bar graph. For example, a line plot for data about how many buttons students are wearing could look something like this:

```
                X
    X           X           X
    X           X           X
    X           X   X   X           X
    X           X   X   X           X
    X   X   X   X   X   X   X
    0   1   2   3   4   5   6
```

            Number of Buttons

Step 3. Ask students to describe the data. What do they notice about it? For data that have a numerical order (How many buttons do you have today? How many people live in your house? How many months until your birthday?), ask questions like these:

"Are the data spread out or close together? What is the highest and lowest value? Where do most of the data seem to fall? What seems typical or usual for this class?"

For data in categories (What is your favorite book? How do you get to school? What month is your birthday?), ask questions like these: "Which categories have a lot of data? few data? none? Is there a way to categorize the data differently to get other information?"

Step 4. Ask students to interpret and predict. "Why do you think that the data came out this way? Does anything about the data surprise you? Do you think we'd get similar data if we collected them tomorrow? next week? in another class? with adults?"

Step 5. List any new questions. Keep a running list of questions you can use for further data collection and analysis. You may want to ask some of these questions again.

Variations

Data from Home For homework, have students collect data that involve asking questions or making observations at home: What time do your brothers and sisters go to bed? What do you usually eat for breakfast?

Data from Another Class or Other Teachers Depending on your school situation, you may be able to assign students to collect data from other classrooms or other teachers. Students are always interested in surveying others about questions that interest them, such as this one: When you were little, did you like school?

Categories If students take surveys about "favorites"—flavor of ice cream, breakfast

Continued on next page

cereal, book, color—or other data that fall into categories, the graphs are often flat and uninteresting. There is not too much to say, for example, about a graph like this:

```
X
X                       X
X       X       X       X
X       X       X       X               X
X       X       X       X       X       X
vanilla chocolate Rocky chocolate strawberry vanilla
                  Road   chip              fudge
```

It is more interesting for students to group their results into more descriptive categories, so that they can see other things about the data. In this case, even though vanilla seems to be the favorite in the graph above, another way of grouping the data seems to show that flavors with some chocolate in them are really the favorites.

Chocolate flavors   //// //// //

Flavors without chocolate   //// /

The following activities will help ensure that this unit is comprehensible to students who are acquiring English as a second language. The suggested approach is based on *The Natural Approach: Language Acquisition in the Classroom* by Stephen D. Krashen and Tracy D. Terrell (Alemany Press, 1983). The intent is for second-language learners to acquire new vocabulary in an active, meaningful context.

Note that *acquiring* a word is different from *learning* a word. Depending on their level of proficiency, students may be able to comprehend a word upon hearing it during an investigation, without being able to say it. Other students may be able to use the word orally, but not read or write it. The goal is to help students naturally acquire targeted vocabulary at their present level of proficiency.

We suggest using these activities just before the related investigations. The activities can also be led by English-proficient students.

Investigation 1

the numbers 1 to 100

1. Create action commands that ask students to nonverbally identify numbers written on the board or shown on a 100 chart.

   Put your finger on the number 6.
   Cover the number 23.

2. Use classroom items (paper clips, rubber bands) to help students count to 100.

3. Challenge students to write different numbers that you call out.

   Write the number 37.

   Write the number 79.

   Write the number 15.

pattern

1. Give students two colors of interlocking cubes. Show them a pattern of cubes in a line, such as blue, red, red, blue, red, red. Ask students to make their own patterns.

2. Draw or write patterns on the board, identify them, and ask students to continue them.

This is a pattern of shapes; what comes next?

□ △ ○ □ △ ○ □ △ ○

This is a pattern in numbers; what comes next? [*Write the numbers in order, then circle and say aloud each even number, pausing at 8.*]

1 ② 3 ④ 5 ⑥ 7 ⑧ 9 10 11 12 ...

This is another kind of number pattern; what comes next? [*Write the numbers in a column.*]

3
13
23
33
43

Investigation 2

money: coins, cents, nickel, dime, quarter, dollar

1. Use a dollar, quarter, dime, and nickel along with action commands to help familiarize students with these words.

   Put a quarter in your hand.
   Give me a dime.
   Put the nickel under the dollar.
   Put all the coins in a pile.
   Fold the dollar in half.

2. Create action commands that require students to identify coins by their value.

   Take a coin that is worth 5 cents.
   Give me a coin that is worth 10 cents.
   Show me which coin is worth the most money.

same, different, agree, disagree

1. As you show students a penny, ask them to *agree* or *disagree* with your statements by nodding or shaking their heads.

   This is a quarter. Hmm—you disagree that this is a quarter.

   This is a penny. Everyone agrees this is a penny.

   A penny is round—agree or disagree?

2. Now show students a nickel in addition to the penny. Ask them to help you find ways in which

Continued on next page

they are the *same,* and ways in which they are *different.*

Are they both round? Yes, they are the same shape.

Are they both worth 5 cents? No, they are worth different amounts.

Do both have a picture on one side? Is the picture the same or different?

Are they the same color or a different color?

3. Continue asking students to compare other sets of coins, giving statements that they can respond to by agreeing or disagreeing, as well as pointing out attributes that they can identify as being the *same* or *different.*

add, plus, combine, more, less, most, least

1. Give students different numbers of cubes. Ask them to count how many they have. Then combine two students' cubes into one pile. Ask students to calculate the total.

If I combine Khanh's cubes with Liliana's cubes, how many will be in the new pile? If I add Cesar's cubes and Su-Mei's cubes, how many will I have?

2. Have students combine more of their piles to find the new totals. Challenge students to add together the piles of several students. As they do so, write each total on a stick-on note next to the pile, and ask questions that require them to make comparisons.

Look at the totals for these two piles of cubes. Which quantity is more? Which is less?

3. Ask the students to identify which totals represent the most and the least.

subtract, minus, check, double-check

1. Show students a pile of 14 pennies. Slide 6 of these to one side as you explain:

I have 14 pennies. Then I take away, or I subtract, 6 pennies. How many are left?

2. Write $14 - 6 = 8$ on the board, talking through the problem again.

I had 14 pennies. I subtracted 6, and I had 8 pennies left. Fourteen minus 6 equals 8.

3. As you look at the answer, pretend to be unsure. Tell the students that you want to *check* your answer. Take 6 pennies away from the 14. Count aloud the remaining pennies and nod your head.

4. Looking at the answer uncertainly one more time, explain that you are going to *double-check* your answer. Do so with a calculator or on the 100 chart.

5. Continue creating comparing problems in which to compare two amounts, students subtract, then check and double-check their answers.

divide, equal, unequal, group

1. Divide 10 pencils in two equal groups. Identify the piles as equal by showing how each group has an *equal* number of pencils.

2. Divide the pencils into two unequal piles. Shake your head as you explain that these groups have an *unequal* number of pencils.

3. Repeat the same procedure, using other classroom items (books, paper clips).

4. Challenge students to demonstrate comprehension of these words by following action commands.

Divide these pencils into two equal groups.

Divide these paper clips into three equal groups.

Divide these books into four unequal groups.

Blackline Masters

———————————— , 19 ———

Dear Family,

We are beginning a unit called *Mathematical Thinking at Grade 3*. This unit will help your child get used to solving problems that take considerable time, thought, and discussion. While solving these problems, your child will be using materials like pattern blocks, cubes, and calculators, and will be writing, drawing, and talking about how to do the problems. Emphasis during this unit will be on thinking hard and reasoning carefully to solve mathematical problems.

During this unit your child will explore even and odd numbers, create symmetrical designs, look for number patterns, and combine and compare different amounts of money or handfuls of objects. The class will also collect and organize information about themselves as a group—we call this *working with data*. Your child will have a math folder or journal for keeping track of work.

While our class is working on this unit and throughout the year, you can help in several ways:

■ Your child will have assignments to work on at home. Sometimes he or she will involve your participation. For example, your child will be teaching you a game called Plus–Minus–Stay the Same. Later in the unit, your child will be figuring out ways to make $1.00 and will be asking to count the change in your pocket or purse.

■ Often children will work out number problems by using real objects. So, when they are working at home, it would help them to have a large collection of objects for counting, such as beans, buttons, or pennies.

■ In class, students will be making a set of Addition Cards. You will probably recognize these as the addition facts, although we call them combinations. Your child will be sorting these cards into "the ones I know" and "the ones I am working on." Speed is not the goal: The goal is for each child to develop effective strategies for combining numbers. Sometimes for homework, children will choose a few combinations that they are working on and think about strategies that will help them remember them. For example, one strategy a child might use is this:

> What's 6 + 7? Well, I know 6 + 6 is 12, and 6 + 7 is one more than that, so it's 13.

You can help with these combinations by listening to your child's strategies or sharing ones that you use.

We are looking forward to an exciting few weeks as we create a mathematical community in our classroom.

Sincerely,

100 Cubes

I made a group of 100 cubes by _____

_____

_____

_____

Make a record using words, pictures, or numbers
to show how you know you have exactly 100 cubes.

100 Objects

Think of two different ways that you could organize or arrange your objects using words, pictures, or numbers to easily show that you have exactly 100 objects. Show these ways below or on the back of this sheet. Remember to use a sensible but different strategy from the one you used in class for at least one way!

Hint: Think about what we did in class with cubes to give you some clues about organizing your objects. Fill in the name of the objects you counted on the line below.

I used _____ to make groups of 100.

Two Ways I Organized My 100 Objects

Partially Filled 100 Chart

1	2			5		7			10
11					16				
			24					29	
	32						38		
41							48		
		53		55					60
		63				67			
	72		74		76			79	80
81							88		
			95		97				100

Numbers for the 100 Chart

1. Add these numbers to your 100 chart:

26	36	62	8	85
	91	70	57	43
13	83	34	46	19
	14	40	35	69

2. Fill in the rest of your 100 chart.

3. When you are finished, find a way to double-check your chart.

Materials
100 chart for each player
Deck of Numeral Cards
Colored pencils, crayons, or markers (or, if you want to reuse the 100 chart, pennies, buttons, or bits of paper to cover chart squares)

Players: 2

How to Play
1. Decide who will go first. The first player chooses two Numeral Cards from the deck to get a base number. The first card is the tens digit, the second is the ones digit. A Wild Card can be used as any numeral.

2. Decide whether you want to **add** 10 to this number, **subtract** 10 from this number, or **stay** with this number. Cover the resulting number on your 100 chart by coloring it in.

3. The other player now chooses two Numeral Cards from the deck, determines the number, and decides whether to **add** 10 to that number, **subtract** 10 from that number, or **stay** with that number.

4. Put the cards you use in a discard pile. (If you run out of cards, mix these up and use them again.)

5. The goal is to cover five numbers on your 100 chart in a row—across, up and down, or diagonally—before your partner does. The game continues until one player has five in a row.

Half and Half

Side A

Side B

How Many Blocks?

Use this sheet with the Half and Half sheet and your pattern blocks.

Pattern Block Design 1

1. Build Side A of your design.

2. Number of blocks in Side A: _____

3. I think there will be _____ blocks in the whole design because

_____

_____

_____

4. Build Side B to finish your design.

5. How many blocks are in your finished design? _____

Pattern Block Design 2

1. Build Side A of your design.

2. Number of blocks in Side A: _____

3. I think there will be _____ blocks in the whole design because

_____

_____

_____

4. Build Side B to finish your design.

5. How many blocks are in your finished design? _____

Strategies for Addition

Write down two addition combinations
you are trying to learn better.
Write down a strategy for each one.
Here is an example.

I need to work on _____5_____ + _____4_____ .

Here is a strategy that helps: _I think of 5+5 because I_
know that 5+5=10. And 5+4 is 1 less, so it's 9.

1. I need to work on _____ + _____ .

Here is a strategy that helps: _____

_____

_____

_____

2. I need to work on _____ + _____ .

Here is a strategy that helps: _____

_____

_____

_____

Problem Strategies

Which problem card did you do? _____

This is what I had to find out (describe the problem
in your own words):

_____

_____

_____

This is how I did it:

Money Problems to Do at Home

quarter

dime

nickel

penny

If you can, borrow some coins from someone at home to help you do these problems.

1. What coins could you have to make exactly $1.00? Find at least three different ways to make $1.00 with coins. Show them with pictures or numbers on the back of this paper. If you can think of more ways, draw those, too.

2. If you have 1 quarter, 6 dimes, 3 nickels, and 4 pennies, how much money do you have? How did you figure out this problem?

3. Find an adult in your house who has some coins in a pocket or purse. Together, figure out how much money that person has altogether.

How Much Is Your Symmetrical Design Worth?

Using coins from home, build a design with mirror symmetry on Student Sheet 5, Half and Half. Then answer the following questions. Feel free to count the objects!

1. Build side A of your design on one side of Student Sheet 5, Half and Half.

2. Amount of money side A is worth: _____

3. I think the whole design will be worth _____ because

   _____

   _____

4. Build side B to finish your design (remember to make it symmetrical).

5. How much is your whole design worth? How did you figure this out?

   _____

   _____

Challenge Questions (You can use the back of the sheet to show your work.)

1. Can you make a symmetrical design worth $1.30? $1.50? $.70? How much would half of each of these designs be worth? How do you know?

2. Create a symmetrical design worth $1.00 using the most coins you can. Now create one using the fewest coins you can.

Addition in Two Ways

Solve this problem in two different ways, and write about how you solved it:

$$25 + 27 =$$

Here is the first way I solved it:

Here is the second way I solved it:

More Addition in Two Ways

Make up an addition problem that you think is
not too easy and not too hard for you. Solve
the problem in two different ways and write
about how you solved it.

The problem I picked is: _____ + _____ =

Here is the first way I solved it:

Here is the second way I solved it:

4 + 3 = 3 + 4 = Clue: _____	5 + 3 = 3 + 5 = Clue: _____
6 + 3 = 3 + 6 = Clue: _____	7 + 3 = 3 + 7 = Clue: _____
8 + 3 = 3 + 8 = Clue: _____	9 + 3 = 3 + 9 = Clue: _____
10 + 3 = 3 + 10 = Clue: _____	5 + 4 = 4 + 5 = Clue: _____
6 + 4 = 4 + 6 = Clue: _____	7 + 4 = 4 + 7 = Clue: _____
8 + 4 = 4 + 8 = Clue: _____	9 + 4 = 4 + 9 = Clue: _____

Investigation 2 • Resource
Mathematical Thinking at Grade 3

10 + 4 = 4 + 10 = Clue: _____	6 + 5 = 5 + 6 = Clue: _____
7 + 5 = 5 + 7 = Clue: _____	8 + 5 = 5 + 8 = Clue: _____
9 + 5 = 5 + 9 = Clue: _____	10 + 5 = 5 + 10 = Clue: _____
7 + 6 = 6 + 7 = Clue: _____	8 + 6 = 6 + 8 = Clue: _____
9 + 6 = 6 + 9 = Clue: _____	10 + 6 = 6 + 10 = Clue: _____
8 + 7 = 7 + 8 = Clue: _____	9 + 7 = 7 + 9 = Clue: _____

10 + 7 = 7 + 10 = Clue: _____	9 + 8 = 8 + 9 = Clue: _____
10 + 8 = 8 + 10 = Clue: _____	10 + 9 = 9 + 10 = Clue: _____
3 + 3 = Clue: _____	4 + 4 = Clue: _____
5 + 5 = Clue: _____	6 + 6 = Clue: _____
7 + 7 = Clue: _____	8 + 8 = Clue: _____
9 + 9 = Clue: _____	10 + 10 = Clue: _____

Doubles and Halves Problem 1

There are 32 children in a class. Divide the class into two teams. Can you make two equal teams?

two teams

Doubles and Halves Problem 2

There are 27 children in a class. They line up in two lines. Can they make two equal lines?

children in line

Doubles and Halves Problem 3

We work at the computer with partners. There are 24 of us. Can everyone have a partner?

partners

Doubles and Halves Problem 4

One cup holds 15 pencils. Another cup holds the same number of pencils. How many pencils are there?

pencils

Doubles and Halves Problem 5

We had a class picnic. I brought 17 cups. My friend brought 17 cups, too. How many cups did we have?

cups

Doubles and Halves Problem 6

There are 49 people on the bus. Each seat holds 2 people. How many seats are filled?

bus seat

Doubles and Halves Problem 7

I have 25 cherries.
I will share them with you.
Can we each have the same
number of cherries?

cherries

Doubles and Halves Problem 8

I have 33 balloons.
I need to make two bunches.
Can I make two equal bunches
of balloons?

balloons

Doubles and Halves Problem 9

There are two fish tanks.
One tank has 23 fish.
The other tank has the
same number of fish.
How many fish
are there
altogether?

fish tank

Doubles and Halves Problem 10

I have 21 apples.
I plan to cut each apple in half.
How many pieces of apple

apple

Doubles and Halves Problem 11

I have 38 flowers in my garden.
My aunt has twice as many
flowers. How many flowers
does she have?

flowers

Doubles and Halves Problem 12

I saw 29 birds sitting on a fence.
How many feet did all those
birds have?

bird

Name _____

Activity Choice   ✔ **When Finished**

1. _____   ☐

2. _____   ☐

3. _____   ☐

4. _____   ☐

5. _____   ☐

6. _____   ☐

7. _____   ☐

8. _____   ☐

Money Problem 1

I have 2 dimes, 1 nickel, and 2 pennies in one pocket.
I have exactly the same coins in my other pocket.
How much money do I have?

Money Problem 2

I have some coins in my two pockets.
I have 45¢ in all.
I have 6 nickels in one pocket.
What coins could I have in my other pocket?

Money Problem 3

I have 1 quarter and 3 nickels in one pocket.
I have exactly the same coins in my other pocket.
How much money do I have?

Money Problem 4

I have 2 dimes, 3 pennies, and 1 nickel in one pocket.
I have exactly the same coins in my other pocket.
How much money do I have?

Money Problem 5

I have 82¢. I put half the money in one pocket,
and half the money in the other pocket.
How much money did I put in one pocket?

Money Problem 6

I have 74¢. I put half the money in one pocket,
and half the money in the other pocket.
How much money did I put in one pocket?

Money Problem 7

I have a total of 50¢ in my pockets.
I have 6 nickels in one pocket.
What coins could I have in my other pocket?

Money Problem 8

My sister and I sold lemonade.
We made $1.40.
We want to split the money evenly.
How can we do this?

Money Problem 9

I picked up trash outside my house.
My mother paid me 2¢ for each piece of trash.
I picked up 29 pieces of trash.
How much did I earn?

Money Problem 10

Two girls earned 35¢.
They want to split it evenly.
How much should each girl get?

Guess My Rule

The data from our class are:

Here's a picture or graph to show the data:

Data Analysis

These data were collected from a third grade classroom.

Type of Hair

Curly	//// ///
Straight	//// //// //// //

How many children are in this class? How do you know?

_____

_____

How many more children have straight hair than have curly hair? How did you figure this out?

_____

_____

If there are 25 children in this class and 8 have curly hair, how many have straight hair? How did you figure this out?

_____

_____

Do you think we would get similar data from this class if we collected it again the following week? the following month? Why?

_____

_____

Extra Challenge: Can you come up with another way to represent these data? Show it on the back.

Calendar Math

Find three different ways to make the number that is today's date. For example, you might write the following expressions if today were the 16th.

$$8 + 8 = 16$$
$$17 - 1 = 16$$
$$4 + 4 + 4 + 4 = 16$$

First Way:

Second Way:

Third Way:

Handfuls

Handfuls of _____

Right hand _____   Left hand _____

Compare your handfuls. Use words, pictures,
or numbers to show how you did this.

Combine your handfuls. Use words, pictures,
or numbers to show how you did this.

Handfuls at Home

If you can, find something around your house
you can grab by the handful. You might try
pennies, marbles, building blocks, buttons,
paper clips, or popped popcorn.

Handfuls of _____

Your RIGHT hand   Your LEFT hand

Right   Left

_____   _____

Compare the handfuls for your right hand and
left hand. Which hand has more?
How many more?

Combine your right and your left handfuls.
How many are there altogether?

Use the other side of this sheet to collect data
from one other person in your house.
Compare and combine that person's handfuls.

Is the Sum Odd or Even?

Odd or Even?

_____ + _____ = _____   _____

_____ + _____ = _____   _____

_____ + _____ = _____   _____

_____ + _____ = _____   _____

_____ + _____ = _____   _____

_____ + _____ = _____   _____

_____ + _____ = _____   _____

_____ + _____ = _____   _____

_____ + _____ = _____   _____

_____ + _____ = _____   _____

On the back, write anything you found out about
adding odd and even numbers.

Dividing Apples: No Halves!

Work with a partner on this sheet. Pretend you're dividing apples between the two of you. Find numbers you can split evenly, without any halves.

Number of apples	**Number of apples each person gets**

1. _____   Our answer_____

   The calculator answer_____

2. _____   Our answer_____

   The calculator answer_____

3. _____   Our answer_____

   The calculator answer_____

4. _____   Our answer_____

   The calculator answer_____

5. _____   Our answer_____

   The calculator answer_____

6. _____   Our answer_____

   The calculator answer_____

7. _____   Our answer_____

   The calculator answer_____

8. _____   Our answer_____

   The calculator answer_____

9. _____   Our answer_____

   The calculator answer_____

10. _____   Our answer_____

   The calculator answer_____

Dividing Apples: Using Halves

Work with a partner. Pretend you're sharing apples. Find numbers you can split so that you each get some whole apples **and** half an apple.

Number of apples	**Number of apples each person gets**
1. _____	Our answer _____
	The calculator answer _____
2. _____	Our answer _____
	The calculator answer _____
3. _____	Our answer _____
	The calculator answer _____
4. _____	Our answer _____
	The calculator answer _____
5. _____	Our answer _____
	The calculator answer _____
6. _____	Our answer _____
	The calculator answer _____
7. _____	Our answer _____
	The calculator answer _____
8. _____	Our answer _____
	The calculator answer _____
9. _____	Our answer _____
	The calculator answer _____
10. _____	Our answer _____
	The calculator answer _____

Odd and Even Numbers

Here are two things I have learned about
odd and even numbers:

Bring this sheet back to class tomorrow!

Writing About Odd and Even Numbers

Write about one idea you think is true about odd and even numbers. Include these things in your writing:

Write down your idea.

Tell WHY you think your idea is true.

Give at least TWO examples of your idea.

Use words, numbers, and pictures to help you explain your idea.

1	2	3	4	5	6	7	8	9	10
11	12	13	14	15	16	17	18	19	20
21	22	23	24	25	26	27	28	29	30
31	32	33	34	35	36	37	38	39	40
41	42	43	44	45	46	47	48	49	50
51	52	53	54	55	56	57	58	59	60
61	62	63	64	65	66	67	68	69	70
71	72	73	74	75	76	77	78	79	80
81	82	83	84	85	86	87	88	89	90
91	92	93	94	95	96	97	98	99	100

0	0	1	1
0	0	1	1
2	2	3	3
2	2	3	3

4	4	5	5
4	4	5	5
6	6	7	7
6	6	7	7

8	8	<u>9</u>	<u>9</u>
8	8	<u>9</u>	<u>9</u>
WILD CARD	**WILD CARD**		
WILD CARD	**WILD CARD**		

Practice Pages

This optional section provides homework ideas for teachers who want or need to give more homework than is assigned to accompany the activities in this unit. The problems included here provide additional practice in learning about number relationships and in solving computation and number problems. For number units, you may want to use some of these if your students need more work in these areas or if you want to assign daily homework. For other units, you can use these problems so that students can continue to work on developing number and computation sense while they are focusing on other mathematical content in class. We recommend that you introduce activities in class before assigning related problems for homework.

Addition in Two Ways   Solving problems in two ways is emphasized throughout this level of *Investigations*. Here we provide three sheets of addition problems that students solve in two different ways. You can make up other problems in this format, using numbers that are appropriate for your students. Students describe each way they solved the problem. We recommend that you give students an opportunity to share a variety of strategies for solving addition problems before you assign this homework.

Doubles and Halves   This type of problem is introduced in this unit. Here you are provided three of these problems for student homework. You can make up other problems in this format, using numbers that are appropriate for your students. Students record their strategies for solving the problems, using numbers, words, or pictures.

Money Problems   This type of problem is introduced in this unit. Here you are provided three of these problems for student homework. You can make up other problems in this format, using numbers that are appropriate for students. Students record their strategies for solving the problems, using numbers, words, or pictures.

Practice Page A

Solve this problem in two different ways, and write about how you solved it:

15 + 30 =

Here is the first way I solved it:

Here is the second way I solved it:

Practice Page B

Solve this problem in two different ways, and write about how you solved it:

19 + 32 =

Here is the first way I solved it:

Here is the second way I solved it:

Practice Page C

Solve this problem in two different ways, and write about how you solved it:

16 + 21 =

Here is the first way I solved it:

Here is the second way I solved it:

Practice Page D

There are 13 dogs in the kennel. How many eyes are in the kennel?

Show how you solved this problem. You can use numbers, words, or pictures.

Practice Page E

I have a bag with 34 carrots. I will share them with you. Can we each have the same number of carrots?

Show how you solved this problem. You can use numbers, words, or pictures.

Practice Page F

We had a picnic at the beach. I brought 18 mini sandwiches. My friend brought 18 mini sandwiches, too. How many mini sandwiches did we have?

Show how you solved this problem. You can use numbers, words, or pictures.

Practice Page G

I have a total of 45¢ in my pockets. I have 2 dimes, 1 nickel, and 2 pennies in one pocket. How much do I have in the other pocket?

Show how you solved this problem. You can use numbers, words, or pictures.

Practice Page H

I have a total of 55¢ in my pockets. I have 3 dimes,
3 nickels, and 3 pennies in one pocket. How much
do I have in the other pocket?

Show how you solved this problem. You can use
numbers, words, or pictures.

Practice Page 1

I have 3 dimes, 1 nickel, and 4 pennies in one pocket. I have exactly the same coins in my other pocket. How much money do I have?

Show how you solved this problem. You can use numbers, words, or pictures.